HANDBOOK OF
HUMAN EMBRYOLOGY

A

HANDBOOK OF

HUMAN EMBRYOLOGY

R. W. HAINES
M.B., D.Sc.

Medical School, Makerere, Kampala
Late of College of Medicine, University of Lagos
and
Medical College, Baghdad

AND

A. MOHIUDDIN
M.B., Ph.D.

Universities of Lagos (Nigeria)
and
Riyadh (Saudi Arabia)

FIFTH EDITION

CHURCHILL LIVINGSTONE
EDINBURGH AND LONDON
1972

First edition	.	.	.	1961
Second edition	.	.	.	1962
Third edition	.	.	.	1965
Fourth edition	.	.	.	1968
Fourth edition reprinted	.	.	1970
Fifth edition	.	.	.	1972

ISBN 0 443 00957 0

Printed in Great Britain

Preface to the Fifth Edition

The authors, who have worked together in London, Baghdad and Lagos, are still wandering round the world in the wrack of the Pax Britannica, one now in Riyadh, the other in Accra. They are still remote from large libraries and would always welcome information or advice.

One of the criticisms we have heard is that our presentation and language are over-simplified. As one embryologist put it, 'It's all right if you like reading the Daily Mirror'. But we are not much impressed by mystiques of any kind (The 'Personal View' of embryology of one of the authors will be found in the B.M.J. March 11th, 1972) and this book was first written for students using English as a second language (and very good some of them have turned out to be). We have tried to be clear, and we hope not at the expense of accuracy.

The puzzled newt on the cover does not represent, as some suppose, the authors cocking a snook at more solemn colleagues: the explanation is given on p. 61. The subject is difficult and a student may still have to decide 'In my mind I'll file and docket Rathke's pouch and Seessel's pocket', though nowadays most of us have disposed of the pocket. He may also feel that 'after June, with any luck, I'll rid my head of all such muck', though it is a little more relevant to his work than it used to be. Of course if someone finds a use for Seessel's pocket it will have to go back.

This edition continues the trend of the last towards emphasis on embryonic abnormalities of clinical interest, particularly where they throw light on normal development. Some material normally covered in the histology course has been omitted, and several new drawings, prepared by Mr. R. Callander from our pencilled sketches, have been added.

R. Wheeler Haines
A. Mohiuddin

1972

From the Preface to the Third Edition

This little book started life in Baghdad, Iraq, as a text which was meant to help medical students grasp the essentials of embryology. The first edition comprised only Part I, covering development up to the 7 mm. stage, the second included a revised Part I and a Part II covering organogeny. Both editions were prepared with the help of the Faculty of Medicine, Baghdad. Dr. Faysal S. Nashat, Editor of the Journal of the Faculty, was entirely responsible for the production of the work and spent many hours sitting beside the block-maker scratching the clichés with a knife to delete unwanted metal.

The present version of the book is based on the Baghdad productions and we again find pleasure in acknowledging our debt to the Faculty of Medicine, Baghdad, and the friends who helped us. We have revised the book and have re-written certain chapters in full or in part so as to bring the text up-to-date. For doing this we have drawn freely upon such works as were available to us; these have been given at the end of the text, under 'Note on Sources'. The general plan of the book, partly 'systematic' and partly 'regional' in treatment of the subject, has been preserved, as we have found this helpful to the students. Dates and measurements have been omitted as a rule because they only distract attention from the sequence of changes. We expect students to refer to standard texts available in medical libraries when information beyond that given in this book is required.

The course suggested can be covered in about 40 hours, including 15 lectures to the 7 mm. stage, 4 practical sessions of $2\frac{1}{2}$ hours each on the 7 mm stage, and a senior course of a further 15 lectures, given after the parts have been dissected in the adult. We prefer the careful study of transverse sections of the 7 mm. pig embryo, rather than a study of odd sections at several stages, but these can be added if time allows. Again, we prefer to use one embryo with the serial sections mounted singly, each on a numbered slide, rather than using several embryos with the sections mounted in rows. One 7 mm. embryo cut at 10 microns gives about 700 slides, enough for a large class to study each region. Loss of an odd slide here and there does not impair the use of the serial. A second embryo should be cut sagittally.

R. WHEELER HAINES
A. MOHIUDDIN

1965

Contents

PREFACE

FIRST PART TO THE 7 mm. STAGE

1 OVUM TO THE TWO-LAYERED STAGE 3

Spermatozoon, ovum, zygote, morula, embedding, primary mesoderm, ectodermal and endodermal vesicles, chorion and amnion, early trophoblast.

2 THE THREE-LAYERED PRESOMITE EMBRYO 15

Primitive streak, notochordal process, three layered embryo neurenteric canal, intraembryonic coelom, neural folds, neural tube, somites.

3 THE FOLDED EMBRYO 27

Folding off, stomatodaeum, Rathke's pouch, reversal, cloaca, blood islands, jelly space, mesenchyme.

4 THE EARLY BLOOD VESSELS AND HEART 33

Angioblasts, vitelline and umbilical circulations, embryonic vessels, coelomic divisions, Y-shaped heart, primary chambers, chest wall, mesocardium, free heart, S-shaped heart, interventricular foramen, division of heart.

5 THE PHARYNX 47

Optic cup, lens, otocyst, branchial grooves and arches, branchial pouches, thyroid and pulmonary diverticula, separation of trachea, aortic arches.

6 THE BRAIN AND SPINAL NERVES 54

Fore-, mid- and hindbrain, neural crest, sensory ganglia, spinal cord, layers of cord, roots of spinal nerves, cranial nerves, control of growth, peripheral nerves.

7 THE LIVER AND STOMACH 62

Hepatic diverticulum, trabeculae, veins of liver, primary loop of gut, mesentery, stomach, lesser sac, pancreatic buds, ductus venosus, sinus venosus.

8 MESONEPHROS AND LIMB BUDS 71

Intermediate cell mass, mesonephric duct, mesonephros, inferior vena cava, cloaca, metanephric bud, limb buds, determination.

9 SECTIONS OF 6 AND 7 mm. PIG EMBRYOS (30 DAYS) . . 79

SECOND PART 7 mm. STAGE TO BIRTH

10 FACE AND NOSE 103

Early face, nasal placodes and cavities, nasal fin, hare lip, mid-line anomalies, nerves of face, nasal septum.

11 THE MOUTH 110

Palatal processes, cleft palate, choanae, lips and gums, teeth, tongue, salivary glands.

12 BRANCHIAL ARCHES: THEIR VESSELS, SKELETON, MUSCLES
   AND NERVES 118
   Arterial arches, division of truncus, aorta, branchial skeleton and
   musculature, recurrent nerves.

13 PHARYNGEAL DERIVATIVES 125
   Cervical sinus, pharyngo-tympanic tube, thyroid gland, parathyroids, thymus, laryngo-pharyngeal orifice, trachea, lung.

14 THE LATER HEART 134
   Superior vena cava, septum secundum, bulbar ridges, aortic and
   pulmonary valves, auriculo-ventricular valves, expansion of auricles,
   conducting system, foetal circulation, changes at birth, coarctation,
   tetralogy of Fallot, coronary arteries, lymphatics.

15 THE PLACENTA AND TWINNING 147
   Expansion of amnion, umbilical cord, placenta, ectopic pregnancy,
   placenta praevia, twins, conjoined twins.

16 THE ABDOMEN 159
   Greater omentum, portal vein, pancreas, liver lobules, lesser omentum, pleuroperitoneal membranes, intestines, Meckel's diverticulum.

17 THE SUPRARENAL, KIDNEY AND CLOACA 174
   Suprarenal, renal tubules, migration of kidney, horse-shoe kidney,
   cloacal septum, vas deferens. Müllerian ducts, uterus and vagina,
   umbilical region, liver attachments, umbilical anomalies.

18 THE GENITAL ORGANS 189
   Gonads, epididymis, genital swellings, penis, vaginal process,
   gubernaculum, broad ligament, breast.

19 THE SKELETON 197
   Mesenchymal skeleton, interzones, primary centres, phocomelia,
   chondrocranium, skull at birth, achondroplasia.

20 THE TRUNK AND LIMBS 204
   Myotomes, sclerotomes, vertebrae, spina bifida, muscles and tendons,
   bursae, limb arteries.

21 THE NERVOUS SYSTEM 214
   Sympathetic system, spinal cord, brain, choroid plexuses, internal
   capsule, commissures, cranial nerve components, cerebellum, cerebrospinal fluid.

22 THE MENINGES, EYE, EAR AND SKIN 227
   Subarachnoid and subdural spaces, eye, anencephaly, cyclopia,
   inner and middle ear, nails and hairs.

23 CONTROL OF GROWTH 236
   Induction, organization, inhibition, hormone, teratomas, function,
   maturity.

   NOTES ON SOURCES 243

   INDEX 247

FIRST PART

TO THE 7mm. STAGE

Ovum to the Two-layered Stage

Spermatozoon. Fig. 1

The morphological changes which a spermatid passes through in order to become a spermatozoon, viz., condensation of nucleus, reduction of cytoplasm and formation of a tail including a flagellum, are all designed to make it motile. The head is formed of nuclear material with a thin covering of cytoplasm. It is attached by a neck to a body which contains

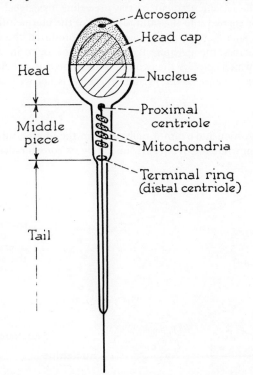

Fig. 1.1. A spermatozoon viewed from one of the broad sides of its flat head.

two centrioles. The anterior centriole gives rise to the flagellum which is continued through the ring-shaped posterior centriole into the tail. The spermatozoon is a very slender cell, about 60 microns long. Although this is about 1/3 of the diameter of the ovum, the mass of the spermatozoon is about 50,000 times smaller than that of the ovum. Most of the length of the spermatozoon lies in the tail, which is capable of undulation. This results in propulsion of the spermatozoon, with the head leading, at the rate of about 1 mm. per minute.

Ovum. Fig. 2

The mature ovum is a large, spherical cell about 1/6 mm. in diameter, *i.e.*, just large enough to be seen by the naked eye against a dark background. It has a large, pale staining nucleus but no centrosome. The cytoplasm contains a small amount of yolk which is evenly dispersed. A zona pellucida and corona radiata enclose the ovum and polar bodies. Usually only one ovum is shed during each menstrual cycle, about 14 days after the commencement of the preceding menstrual flow. The shed ovum is carried into the infundibulum of the uterine tube, usually of the same side as that of the ovary which produced the ovum, but sometimes across the peritoneal cavity into the opposite tube. The mechanism which conveys the ovum into the tube, and not elsewhere in the enormous expanse of the peritoneal cavity, is not well understood.

Fertilization

The spermatozoon and ovum usually meet in the uterine tube near the infundibulum while the ovum surrounded by its corona radiata is being

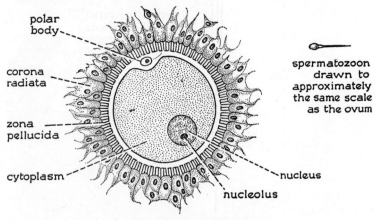

Fig. 1.2. An ovum with its corona.

carried towards the uterine cavity by ciliary and peristaltic action in the tube during the first day after ovulation. If spermatozoa in the normal number of 3–4 hundred million have been deposited in the vagina no more than a day earlier, a sufficient number of them swim into each tube to produce enough hyaluronidase to disperse the corona and allow the head and body/middlepiece of one spermatozoon to enter (the tail breaks off) the cytoplasm of the ovum, after which no more are allowed in. The vitality of spermatozoa and ovum does not last more than 24 hours after discharge, and the risk of foetal abnormalities probably increases with delay in fertilization, if it occurs at all beyond that period. Spermatozoa can, however, be preserved for long periods by quick freezing in a suitable medium. This stops their movement but they can be reactivated by judiciously applied warmth and can fertilize an ovum.

The Zygote. *Fig. 3*

When discharged, the spermatozoa are mature gametes, the haploid karyotypes 22X and 22Y being present in equal strength, though the latter type fertilize the ovum in about 55% of the cases. The ovum completes the second stage of meiosis after penetration by the spermatozoon but retains the bulk of the cytoplasm, and its nucleus becomes the female pronucleus. The head of the spermatozoon swells to form the

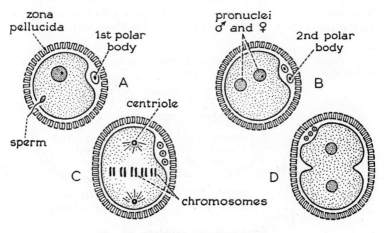

Fig. 1.3. Fertilization and first cleavage.

male pronucleus, while its body provides the centrioles needed for spindle formation in the first mitotic division of the zygote. The karyotype of the ovum is always 22X, and when nuclear envelopes disappear the haploid complements of the parents of the person come together (fusion of pronuclei) to form the normal diploid somatic karyotype of either 44XY (male) or 44XX (female).

Genetic Defects

An increasing number of genetic defects which influence development, both early and late, is being recognized. Defects caused by abnormal genes located on chromosomes of normal appearance, or those due to partial damage to chromosomes, as in translocations, deletions, or ring shapes, are beyond the scope of this text. However, anomalous development due to loss or gain in the number of chromosomes is relevant to us. In most cases this occurs through unequal sharing of chromosomes during meiosis leading to the formation of gametes, although the (exceedingly rare) development of one normal and one mongoloid child from a single ovum shows that unequal sharing may occur during the first few mitoses of the zygote. If it occurs later, apparently normal persons with mosaic patterns of karyotype in their tissues or grossly abnormal chimaerae may develop. Certain karyotypes in the zygote result in foetal death and abortion. These are: loss of even a single complete autosome, absence of X chromosome in the male, general polyploidy, and there may be others. Several kinds of anomalies, including mongolism, result from the presence of an extra autosome, depending on the particular autosome. Females with but a single X chromosome exhibit Turner's syndrome, while those with more than

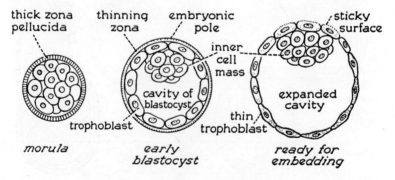

thick zona pellucida

thinning zona

embryonic pole

sticky surface

inner cell mass

cavity of blastocyst

expanded cavity

trophoblast

thin trophoblast

morula

early blastocyst

ready for embedding

Fig. 1.4. Morula and blastocyst.

two X chromosomes look normal but are mentally deficient. The presence of a Y chromosome makes the person male, but if he has more than one X chromosome he exhibits Klinefelter's syndrome, and mental deficiency as well if he has more than two X. Some males with violent tendencies in sex and crime have been known to possess more than one Y chromosome, though the body develops normally.

Cleavage and Formation of Morula. Figs. 3 & 4

Mitosis of the zygote produces 16 cells after four sets of divisions. Multiplication of nuclei at the expense of cytoplasm increases the proportion of nuclear material. The yolk stored in the cytoplasm of the ovum is consumed for production of energy during this period and this causes a reduction in the total mass of protoplasm in the 16 cells as compared with the mass of the zygote, and all the cells are contained within the unbroken zona pellucida. While undergoing division, the zygote is conveyed along the uterine tube towards the uterine cavity and reaches it 3 to 4 days after fertilization. It is now a morula, and spends a further 3 or 4 days free in the upper part of the uterine cavity before it is embedded.

Formation of Blastocyst (Trophoblastic Vesicle). Fig. 4

The cells of the morula continue to divide rapidly and the growing mass draws nourishment and oxygen from the uterine secretions in

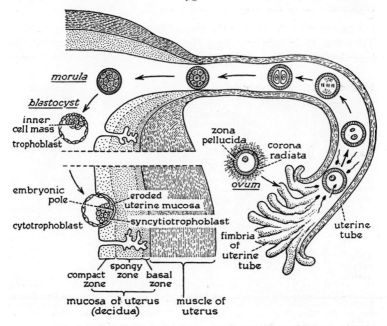

Fig. 1.5. Passage through the uterine tube and embedding.

which it is bathed. Fluid is absorbed in quantities larger than those required in building new cells. The excess of fluid collects between the cells in the form of a pool, enclosed by a single layer of thin flattened cells, the trophoblast. A knob or lump of cells, the inner cell mass, projects into this pool from the trophoblast over a comparatively small area. The entire structure is the blastocyst. It is spherical in form and its outer shell of trophoblast looks towards the uterine mucosa. The region of the blastocyst which is continuous with the inner cell mass is called its embryonic pole.

Positioning of the Embryo. Fig. 5

As the blastocyst expands the zona pellucida thins and finally disappears, so that the trophoblast comes into direct contact with the uterine mucosa. The outer surface of the trophoblast at the embryonic pole is more sticky than the rest, and it is normally this surface that first becomes attached to the mucosa and leads the way in embedding. The blastocyst usually finds a site in the upper part of the uterine cavity and seems to avoid the openings of the uterine glands.

Differentiation of the Trophoblast. Figs. 5 & 6

On contact with the mucosa the trophoblastic cells become very active. They lose their flattened shape, becoming plump and rounded. An inner layer forms a continuous epithelium of cubical type with distinct

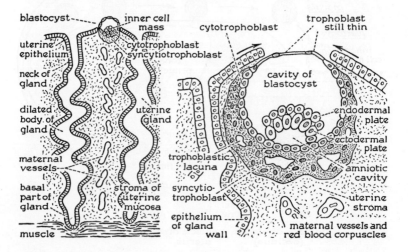

Fig. 1.6. Embedding.

cells, the cytotrophoblast. An outer layer, the syncytiotrophoblast, is thick and spongy, with cavities, the trophoblastic lacunae, scattered through its substance. There are no cell boundaries so that the numerous nuclei are scattered in the cytoplasm.

Embedding of the Embryo. Figs. 5 & 6

Just before embedding the metabolic rate of the embryo is increased many-fold, with the formation of bicarbonates which pass into the adjacent maternal mucosa. At the same time, under the influence of progesterone circulating in the maternal blood, carbonic anhydrase increases in the mucosa. This enzyme dissociates the bicarbonates, setting free carbon dioxide which is carried off in the blood, and leaving carbonates which accumulate and make the mucosa alkaline. The alkalinity softens the cement substance that holds the mucosal cells together, so that, near the embryo, they fall apart, leaving the embryo embedded in an implantation cavity. The embryo itself is not disintegrated, for its outer layer of syncytiotrophoblast has no cell boundaries and so no cement to soften. However, once past the epithelial barrier, further penetration by the blastocyst occurs through destruction of decidual tissue. Proteolytic enzymes are probably secreted by syncytiotrophoblast until penetration is completed within a week.

Orientation of the Embryo. Fig. 6

Usually the embryonic pole of the blastocyst makes the first attachment and sinks deepest. Where the blastocyst is not yet embedded the cells of the trophoblast remain thin and are not differentiated into cyto- and syncytiotrophoblastic layers. Normally the invasion is confined to the superficial layer of the mucosa between the necks of the glands. When, in exceptional cases, the syncytiotrophoblast is abnormally active and erodes further it may form a particularly virulent cancer, a choriocarcinoma. The embedding process is helped by the reaction of the maternal tissues to the ovum, for the mucosa grows over the surface of the ovum. The point of entry is soon repaired by growth of uterine epithelium without leaving any trace of the site. Further growth of the blastocyst causes a localized dome-like projection of the uterine mucosa into the uterine cavity.

The Embedded Embryo. Fig. 6

Shortly after embedding the cytotrophoblast still forms a simple spherical shell, one cell thick, but the syncytiotrophoblast, now much enlarged to form a spongy mass with trophoblastic lacunae, occupies the greater part of the erosion cavity. The lacunae are filled with the debris of the destroyed uterine mucosa and the glycogen-rich stromal cells, mixed with stagnant blood derived from the vessels of the mucosa opened up in the course of embedding. These, disintegrated by auto-lysis, constitute the embryotrophe, at this time the only source of nutrition for the embryo.

The Germ Layers and Amniotic Cavity. Fig. 6

The inner cell mass divides into a thicker sheet of ectoderm and a thinner sheet of endoderm, the first germ layers. A fluid filled split, the amniotic cavity, appears between the embryonic ectoderm and the inner surface of the trophoblast.

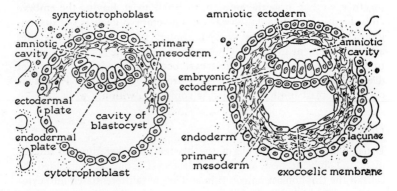

Fig. 1.7. The amniotic cavity and exocoelic membrane.

Primary Mesoderm and Exocoelic (Exocoelomic) Membrane. Fig. 7

Cells migrate from the inner surface of the trophoblast to give a third layer to the blastocyst wall, the primary mesoderm. Histologically this is a loose mesenchyme with comparatively few cells, which are capable of amoeboid movement. The cells lie in a mass of jelly-like intercellular substance and contact each other by means of processes. The cells of the primary mesoderm next the amniotic cavity become specialized to form a low columnar epithelium, the amniotic ectoderm, which stands in sharp contrast to the tall columnar embryonic ectoderm. The cells of the primary mesoderm next the blastocyst cavity become flattened to form a thin continuous sheet of pavement epithelium, the exocoelic membrane.

Ectodermal and Endodermal Vesicles. Fig. 8

As it is formed, the blastocyst cavity is too large at first, so a part of it is pinched off and lost in the primary mesoderm, leaving a remainder of a size comparable to the amniotic cavity. The endoderm grows round the interior of this reduced cavity to complete the endodermal vesicle. The embryo now presents a trophoblast of two layers, cyto- and syncytiotrophoblast, enclosing a mass of primary mesoderm. Embedded in this mesoderm are two vesicles, an ectodermal and an endodermal. The ectodermal vesicle is lined by two kinds of ectoderm, amniotic and embryonic, and encloses the amniotic cavity. The endodermal vesicle is lined by ordinary endoderm throughout. The region that will eventually form the body of the adult is indicated by broken brackets in the diagram. This, the embryonic disc, includes only single layers of ectoderm and endoderm, so that this stage is known as the two-layered embryo. The other parts, not included in the brackets, will form only temporary structures, discarded at the time of birth.

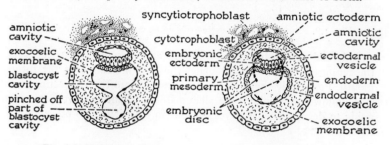

Fig. 1.8. Reduction of the yolk sac and the two-layered embryo.

Extra-embryonic Coelom. Fig. 9

A number of splits appear in the primary mesoderm and soon run together to form the first coelom, known, since it lies outside the embryonic area proper, as the extra-embryonic coelom. At the same time a condensation appears in the primary mesoderm, the connecting stalk, which marks, for the first time, the posterior end of the embryo, as distinct from the anterior end. Since the ventral (endodermal) and dorsal (ectodermal) surfaces are already distinct this implies distinction of the right and left sides.

Prochordal Plate and Allantois. Fig. 9

The coelom spreads, cutting off the two vesicles from the inner surface of the trophoblast except in the region of the connecting stalk which remains as the sole attachment of the vesicles. While this goes on the

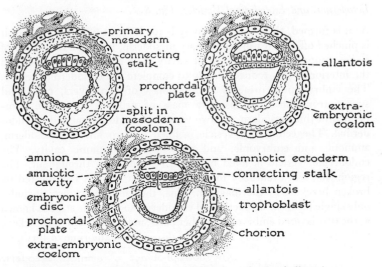

Fig 1.9. Prochordal plate, extra-embryonic coelom and allantois.

endodermal vesicle becomes thickened at two points. Anteriorly the thickening is the prochordal plate, the first indication of the future mouth. Posteriorly the thickening grows out into the connecting stalk as a hollow diverticulum, the allantois. The allantois serves later as a guide to developing blood vessels, but in man it has no other known function.

Chorion, Amnion and Endodermal Vesicle. Fig. 9

The development of the extra-embryonic coelom defines the chorion and amnion. The chorion is the three-layered membrane that encloses the extra-embryonic coelom, consisting of two layers of trophoblast and one of primary mesoderm. The amnion encloses the amniotic cavity and is two-layered, consisting of primary mesoderm and amniotic ectoderm. The part of the endodermal vesicle that projects into the coelom is also covered with primary mesoderm. The two layers of this part, endoderm and primary mesoderm, will later form the yolk sac. The three structures, chorion, amnion and yolk sac, remain separate so long as the extra-embryonic coelom persists.

The Early Trophoblast. Figs. 6 & 7

At the time of embedding the cytotrophoblast was a simple smooth cubical epithelium and the syncytiotrophoblast a massive spongy layer enclosing scattered irregular cavities, the trophoblastic lacunae. The embryotrophe filling these spaces was the result of digestion of the mucosa by enzymes of the trophoblast. But once the embryo is fully embedded no further erosion occurs, enlargement of the space occupied by the embryo being achieved by the growth and displacement of maternal tissues, not by their destruction. Both layers of trophoblast become adapted to a more elaborate form of nutrition of the embryo.

Blood-borne Nutrition. Fig. 10

The small blood vessels of the uterine mucosa now open directly into the trophoblastic lacunae, and these lacunae open into each other so that blood can circulate in the cavities. Here the maternal blood is in direct contact with syncytiotrophoblast and substances needed for the growth of the embryo can be extracted. At this stage the embryo, which till now has changed little in bulk—the morula is actually a little smaller than the fertilized ovum—begins to grow rapidly.

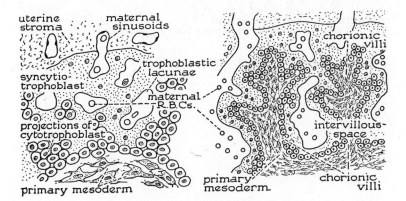

Fig. 1.10. Trophoblast and chorionic villi.

Chorionic Villi. Fig. 10

Projections appear on the outer surface of the cytotrophoblast, and these spread into the syncytiotrophoblast, branching as they go. The primary mesoderm invades these projections, forming their cores. Each projection, with its covering of cyto- and syncytiotrophoblast, is a chorionic villus, and the trophoblastic lacunae, now filled with circulating maternal blood, become the intervillous spaces. The villi are joined to each other and to the walls of the implantation cavity by syncytiotrophoblast.

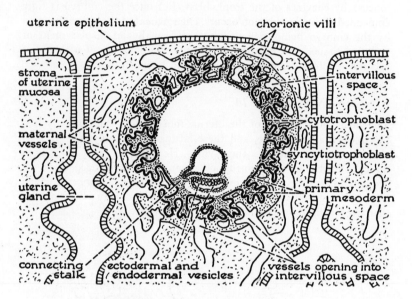

Fig. 1.11. The intervillous circulation.

2

The Three-Layered Presomite Embryo

Embryonic Disc. Figs. 1 & 2

A window cut in the chorion shows the embryo attached by its connecting stalk, but otherwise floating freely in the fluid of the extra-embryonic coelom. The two vesicles concerned with the embryo proper, ectodermal and endodermal, can be made out through the layer of primary mesoderm that covers them, a slight groove on the surface indicating the position of the embryonic disc between them.

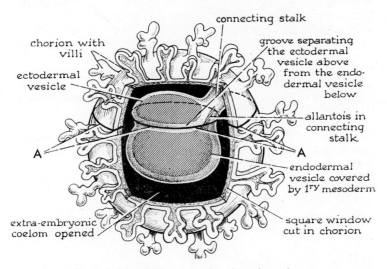

Fig. 2.1. The chorion opened to show the embryo.

A cut through the 'horizontal' plane AA will remove a cap from the chorionic vesicle as the top is removed from an egg, and also cut away a part of the amnion, so as to open the amniotic cavity and expose the embryonic disc in its floor. This cut passes posteriorly through the attachment of the connecting stalk to the vesicle and through the allantois as it lies in the stalk. The disc itself appears, at this stage, as a flat, nearly circular area, covered with ectoderm and continuous with the cut edge of the amnion.

Primitive Streak and Cloacal Membrane. Fig. 2

The ectodermal cells near the midline of the posterior half of the embryonic disc now lose their columnar shape, become many layered and begin to divide rapidly, forming the primitive streak. At the cranial end of the streak a rounded thickening forms the primitive node (or knot). At its posterior end, between it and the connecting stalk the ectoderm also thickens, but here it fuses with the underlying endoderm at the attachment of the allantois to the endodermal vesicle to form the cloacal membrane.

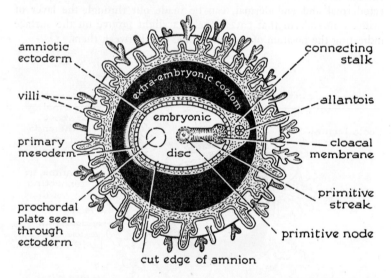

Fig. 2.2. A slice removed along the line AA in Fig. 1.

Notochordal Process. Fig. 3

The embryonic disc is at first circular and the primitive node lies close behind the prochordal plate. But the disc elongates and the node separates from the plate, laying down as it does so a rod of cells, the notochordal process, which projects forwards from the node between the ectoderm and endoderm. A pit develops at the centre of the primitive node, and eventually the process hollows, forming a tubular structure with a lumen, the notochordal canal, opening at the pit into the amniotic cavity.

Three-Layered Embryo. Fig. 4

A transverse section at XX (fig. 3) shows the ectodermal thickening forming the streak, and, at a later stage, a sheet of new mesoderm, the secondary or streak mesoderm on either side, spreading from the primitive streak between the ectoderm and endoderm. Its cells are produced partly by proliferation in the streak itself, and partly by

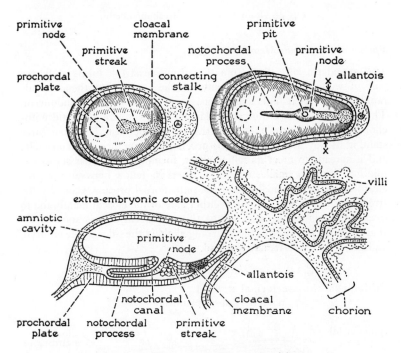

Fig. 2.3. The primitive streak and notochordal process.

migration of cells as indicated by the arrows. Ectodermal cells move towards the midline, are then moved below the surface into the streak and are finally extruded as streak mesoderm. The primary mesoderm, which has hitherto intervened to some extent between the ectodermal and endodermal vesicles, is forced out by the layer of streak mesoderm, which fuses with it at the edge of the disc. The embryo is now in its three-layered stage.

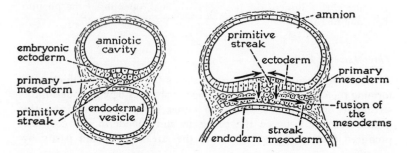

Fig. 2.4. Formation of the streak mesoderm.

Spread of the Streak Mesoderm. Fig. 5

The mesoderm spreading from the primitive streak forms at first a pair of wings lying on either side of the streak, but then spreads rapidly wherever it finds space between the ectoderm and endoderm. Thus it spreads forward on either side of the primitive node and notochordal process, but cannot cross the midline because these structures stand in its way. But cranial to the prochordal plate there is no obstacle, and it spreads to meet its fellow sheet, forming a cross bar across the midline. A much smaller extension meets its fellow between the prochordal plate and cranial end of the notochordal process. Caudally the streak mesoderm spreads round either side of the cloacal membrane to invade the connecting stalk. Thus the mesoderm at this stage is U-shaped, with the limbs of the U arranged along the sides of the embryo, and the cross bar at its anterior end.

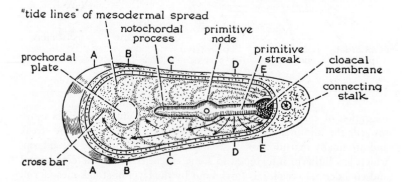

Fig. 2.5. Progressive spread of the mesoderm. (The letters A-E refer to the sections in Fig. 6).

Sections of the Three-Layered Embryo. Fig. 6

A series of diagrammatic sections brings out the structure clearly. Section A (position shown in fig. 5) shows the cross bar of the U-shaped mesoderm lying between the ectoderm and endoderm, with the primary mesoderm on either side. Section B passes through the prochordal plate, with separate sheets of mesoderm on either side. In C the notochordal process with its canal occupies the midline. In D the mesodermal sheets are continuous with the primitive streak from which they have been formed. In E the ectoderm and endoderm are joined to form the cloacal membrane while the mesoderm passes towards the connecting stalk on either side. It may be noted too that as the embryo

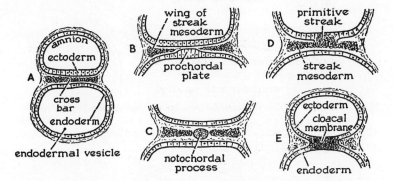

Fig. 2.6. A series of sections cut through a three-layered embryo.

grows, the streak, which once occupied a full half of the length of the embryonic plate, becomes relatively smaller, for once the streak mesoderm is formed its work is done.

Notochordal Plate and Neurenteric Canal. Figs. 7 & 8

We left the notochordal process as a hollow, blind ending tube lying in the midline between the ectoderm and endoderm. The floor of this tube fuses with the underlying endodermal cells and then breaks down, putting the lumen of the tube into continuity with the cavity of the endodermal vesicle. The remainder of the process flattens out to form the notochordal plate, which is continuous with the endoderm on either side. The amniotic and endodermal cavities communicate directly through a narrow canal piercing the primitive node. The communication is called the neurenteric canal because the region of the primitive node is incorporated into the neural tube and the roof of

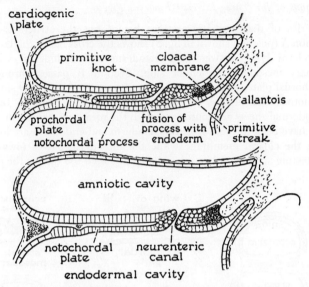

Fig. 2.7. The notochordal process and notochordal plate.

the endodermal vesicle into the intestine at a later stage. Normally the canal disappears completely later on.

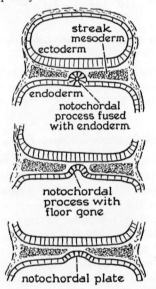

Fig. 2.8. Sections through the notochordal process.

Notochord. Fig. 14

The inclusion of the notochordal tissue as a strip in the roof of the endodermal vesicle is only temporary. Before long it separates again, this time to give a solid rounded cord of cells surrounded by a thin, non-cellular sheath. This is the notochord itself, the first element of the skeletal system.

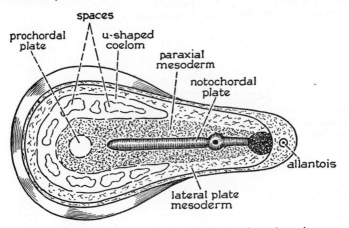

Fig 2.9. Formation and spread of the intraembryonic coelom.

The Intraembryonic Coelom. Fig. 9

A strip of streak mesoderm lies on either side of the notochordal plate forming the paraxial mesoderm, while the mesoderm nearer the edge of the disc is the lateral plate mesoderm. Spaces appear in the lateral plate mesoderm and these soon run together to give a U-shaped coelom, with two limbs lying in the two lateral plates and a cross-piece joining them. The cross-piece is destined to form the pericardial cavity, and the limbs, the pleural and peritoneal cavities.

Communication of Coeloms. Figs. 10, 11 & 12

From the time of its earliest formation the streak mesoderm lies in contact with primary mesoderm at the edges of the disc, and the two mesoderms soon fuse. So the coelomic split in the streak mesoderm can spread laterally and eventually opens into the extraembryonic coelom as indicated by the arrows. At this stage then a coelomic canal traverses the embryo, beginning on one side in the groove between the ectodermal and endodermal vesicles, crossing the midline near the anterior margin of the embryonic plate and ending on the other side.

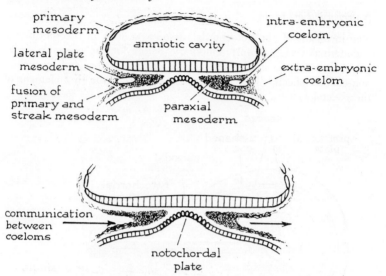

Fig. 2.10. Sections of coelom.

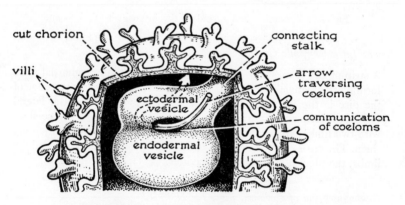

Fig. 2.11. The extra- and intraembryonic coeloms

Neural Plate. Fig. 13

A thickening of the embryonic ectoderm, the neural plate, forms the first rudiment of the nervous system. The plate extends from a point just cranial to the notochordal process to include the primitive node with the neurenteric canal and a part of the primitive streak. It is wider

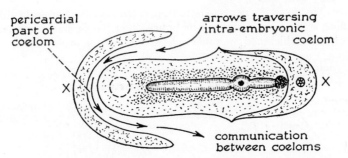

Fig. 2.12. Course of coelom in the embryo.

cranially, where it will form the brain, than caudally, where it will form the spinal cord. It overlies the notochordal process and paraxial mesoderm on either side, but it does not cover the lateral mesoderm.

Neural Folds. Fig. 14

The under surface of the plate grows more quickly than the upper, as indicated by the greater number of mitotic figures. So the edges of the plate bend dorsally, carrying the embryonic ectoderm with them. The two raised ridges so formed are the neural folds. These folds eventually meet and fuse in the midline. The neural plate, which is several cells thick, forms the neural tube, enclosing a large central

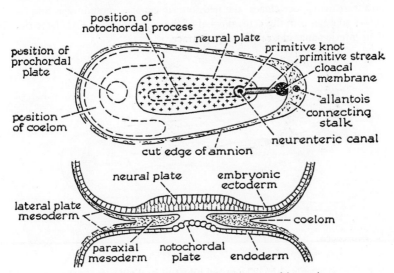

Fig. 2.13. The neural plate in dorsal view and in section.

canal. The much thinner ectoderm from either side meets over the tube and separates from it. It may be noticed that the paraxial mesoderm has become thickened on either side with the rising of the folds.

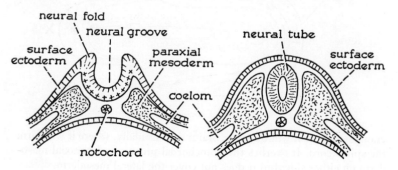

Fig. 2.14. The neural folds and their fusion.

Neural Tube. *Fig. 15*

Closure of the tube begins near its middle, and proceeds gradually towards either end. Cranially and caudally the tube is still open at the anterior and posterior neuropores. The neurenteric canal can be seen through the posterior neuropore. Eventually the neuropores close completely. The region near the anterior neuropore remains as the thin lamina terminalis, which closes the anterior end of the tube. As the posterior neuropore closes, the neurenteric canal is roofed over, and so comes to open into the central canal of the spinal cord. A little later the neurenteric canal closes off without trace.

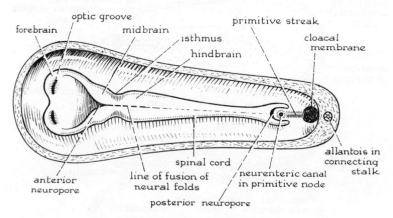

Fig. 2.15. The neural tube and its divisions.

Division of the Neural Tube. Fig. 15

Even while the neuropores are still widely open the main divisions of
the C.N.S. can be made out. The brain takes up more than half the
length of the tube. A pair of bulging masses indicates the fore-brain,
each mass marked internally by a curved optic groove, the first sign of
the later optic outgrowth. A slight swelling indicates the tectum of the
midbrain. A narrowing, the isthmus, separates the mid-brain from the
hindbrain whose walls show characteristic undulations. The hindbrain narrows into the spinal cord.

Somites. Fig. 16

As the neural folds rise the paraxial mesoderm thickens (fig. 10) and
by the time they have closed it forms a large mass lying on either side
of the neural tube and notochord. This mesoderm segments, the more
cranial part first and the more caudal later as the embryo elongates,
forming the somites. The first somite lies near the hindbrain, the others
on either side of the spinal cord, which is, in these early stages, relatively
short. Each somite is separate from its fellows, but is joined laterally to
the unsegmented lateral plate mesoderm. The somites are the first
segmented structures of the body, and impose their segmental pattern
on other structures, the axial skeleton, trunk muscles, trunk vessels and
spinal nerves.

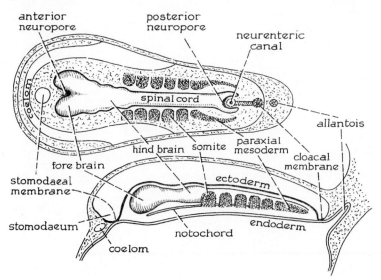

Fig. 2.16. The mesodermal somites and stomodeal membrane.

It may be noted that at this stage the head, distinguished by its inclusion of the brain, is large, and so is the neck, represented by the anterior somites, while the rest of the body, whose paraxial mesoderm is not yet segmented, is relatively short.

Stomodeum and Stomodeal (Stomatodaeal, Oral, Buccopharyngeal) Membrane. Fig. 16

The position of the mouth is marked out early by the prochordal plate thickening of the endoderm (fig. 6). This plate fuses with the overlying ectoderm to form the stomodeal membrane, which then thins. In this region of the membrane, the ectoderm becomes progressively depressed to form the stomodeum, the earliest mouth cavity. It should be noted that the anterior end of the neural tube lies caudal to the stomodeal region, and the anterior end of the notochord a little more caudal still. The embryo now has a notochord, somites, coelom, neural tube with brain and spinal cord, ectodermal covering and endodermal floor, stomodeum and stomodeal membrane, neurenteric canal, remains of the primitive streak and cloacal membrane. It is ready to fold off.

3

The Folded Embryo

Folding Off: General. Fig. 1

So far the embryo has consisted of two rounded vesicles, ectodermal and endodermal, and the associated mesoderm. The ectodermal vesicle now becomes C-shaped, the ends bending towards each other so as to form a head and tail to the embryo. This process pinches off the gut from the yolk sac by a narrow neck, the yolk stalk. When folding off is completed the three regions of the gut, the fore-, mid- and hind-gut, can be distinguished, cranial, opposite and caudal to the yolk stalk with the allantois attached to the hind-gut. The small diagram shows the process in very simplified form.

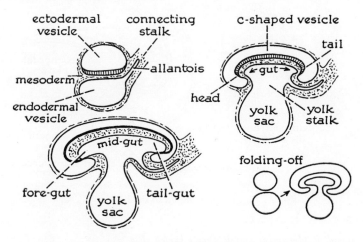

Fig. 3.1. Simplified scheme of folding off in sagittal plane.

The Head End. Fig. 2

The streak mesoderm at the head end of the embryo where it crosses the mid-plane is already differentiated into four parts. Most anteriorly is

27

the septum transversum mesoderm (1) fused with the primary mesoderm covering the ectodermal and endodermal vesicles. Then follows the mesoderm split by the U-shaped coelom where it crosses the midplane (2). Behind this the mesoderm is thin, due to the bulging of the prochordal plate (3), and behind this again the mesoderm is heaped up against the neural tube and notochord (4).

The Stomodeum. Fig. 2

As the embryo folds off the prochordal plate comes to form the cranial end of the fore-gut. The ectoderm sinks towards it, forming the stomodeum. The thin mesoderm in this position is driven out so that the ectoderm and endoderm can come into contact, and fuse to form the stomodeal membrane. The prochordal plate also buds off cells of mesodermal type, the prochordal mesenchyme, which later segments to make the poorly defined head myotomes.

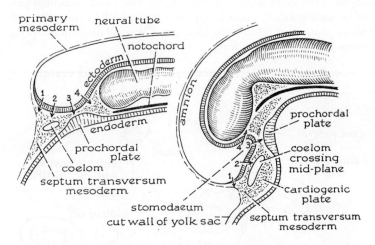

Fig. 3.2. Folding off of the head.

Stomodeal Membrane and Rathke's Pouch. Fig. 3

The stomodeal membrane closes off the stomodeum from the cranial end of the fore-gut which is now expanded to form the pharynx. Just cranial to the membrane the ectoderm forms a blind depression, Rathke's pouch, the first indication of the pituitary gland. This pouch burrows into the mesodermal tissues underlying the forebrain.

Cardiogenic Plate. Figs. 2 & 3

The coelom splits the streak mesoderm into two layers, one next the ectoderm and the other next the endoderm. Where the mesoderm crosses the median plane the layer next the endoderm becomes thickened to form the cardiogenic plate, the first rudiment of the heart. When the process of folding off is complete the cardiogenic plate lies ventral to the pharynx, while the septum transversum lies next the yolk stalk.

Reversal. Figs. 2 & 3

It may be noted that of the four regions marked off earlier (1) gives the septum transversum, (2) the coelom and cardiogenic plate, (3) the stomodeum and buccopharyngeal membrane and (4) the nasal region. During the folding off the order of these structures taken from head to tail has changed from 1-2-3-4 to 4-3-2-1, that is they have been turned through two right angles to give a reversal of the regions. The septum transversum, originally at the extreme anterior end of the embryonic plate, has been tucked into its proper place behind the heart.

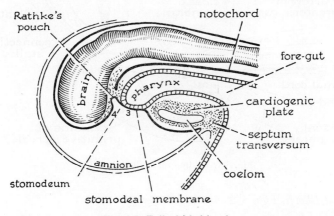

Fig. 3.3. Fully folded head.

Reversal at the Posterior End. Fig. 4

In the flat embryo, before folding off, the primitive streak, cloacal membrane and connecting stalk lie in that order posterior to the primitive node and neurenteric canal. After folding off the node is incorporated in the tail bud and the canal closed off. The bud does not develop far in the human embryo, and soon regresses. The occasional 'tail' of man ('homo caudatus') is a fibro-fatty appendage.

The Cloaca. Fig. 4

The primitive streak is turned onto the ventral surface of the embryo, but as its work of forming streak mesoderm is completed, it degenerates into ordinary connective tissue which is included in the anococcygeal body, while the cells on the surface revert to their original type, viz. ectoderm. The allantois is now attached to the ventral surface of the hind-gut demarcating the hindmost part as the cloaca. The cloacal membrane, lying between the connecting stalk and the remains of the primitive streak, is still thick.

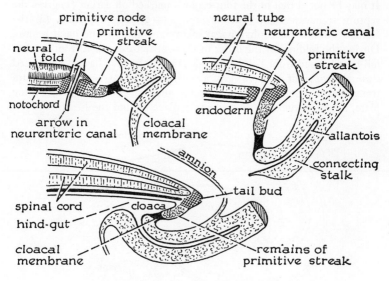

Fig. 3.4. Folding off of the tail end.

Blood Islands. Figs. 5 & 6

As the embryo folds off the vascular system is established. Swellings appear in the primary mesoderm covering the yolk sac, the blood islands. The cells of the islands are at first undifferentiated, but soon the peripheral cells flatten and join edge to edge to form an endothelium, while the cells in the interior of the islands become rounded and develop haemoglobin in their cytoplasm, forming the primitive erythroblasts. A fluid plasma accumulates within the endothelium-lined cavities, probably formed by the endothelium, so that each island consists of a little vesicle of blood ready to join a circulation when this becomes available.

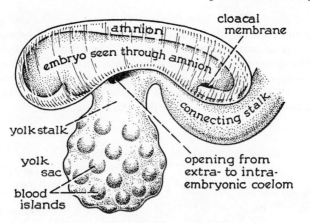

Fig. 3.5. Blood islands on the yolk sac.

Blood Vessels of the Yolk Sac. Fig. 6

The islands become irregular in shape, sending processes which join their fellows to form an anastomosing network of vessels over the surface of the sac. New islands arise as the earlier islands develop into blood vessels so that blood formation is continuous. Later the corpuscles acquire increased amounts of haemoglobin and lose their nuclei to become primitive erythrocytes.

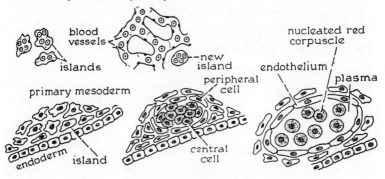

Fig. 3.6. Blood islands and red blood corpuscles.

The Jelly Space. Fig. 7

So far the germ layers, neural tube, notochord, etc. have been considered as if entirely separate from each other, and so, indeed, they appear to be in ordinary stained microscopical preparations. But probing slices of living embryos soon shows that whereas the spaces within the neural tube, the ectodermal and endodermal vesicles and between the two layers of lateral mesoderm are true spaces filled with fluid, the apparent spaces surrounding the neural tube, notochord, etc., are filled with a jelly which holds the embryonic parts together. If the notochord is pushed aside it springs back into place by the resilience of this jelly, which may be described as filling the 'jelly space'.

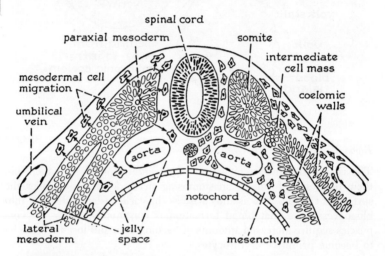

Fig. 3.7. Mesoderm and mesenchyme, left side an earlier stage.

The Mesenchyme. Fig. 7

Mesodermal cells migrate into the jelly, filling up the corners of the embryo, invading the jelly throughout its extent. The cells and jelly together constitute the mesenchyme, an embryonic form of connective tissue, as yet without fibres.

Mesodermal Differentiation. Fig. 7

After the migration of the mesenchymal cells the remainder of the original mesoderm still appears condensed with close-packed cells as opposed to the loosely arranged mesenchyme. Medially the paraxial mesoderm has become segmented to form the somites while the lateral mesoderm has split to form the walls of the coelom. Between these two regions there now lie connecting elements of condensed mesodermal tissue, the intermediate cell masses. These are the first rudiments of the urogenital system.

4

The Early Blood Vessels and Heart

The Angioblasts. Fig. 1

Besides the cells that form the mesenchyme others, the angioblasts,
migrate out from the mesoderm into the jelly space. These align themselves in three main positions: beside the notochord A, between the
mesoderm and ectoderm U, and between the mesoderm and endoderm
V, corresponding to the later positions of the aorta and the umbilical
and vitelline veins.

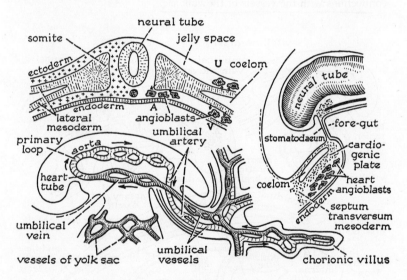

Fig. 4.1. Angioblasts and the earlier blood vessels.

The Heart Angioblasts. Fig. 1

In the cardiac region, where the U-shaped coelom crosses the median
plane, the coelomic wall next the endoderm is already thickened to

c

form the cardiogenic plate. Now angioblasts migrate from the plate into the jelly space so as to lie between the plate and the endoderm. These are the heart angioblasts so that the heart is now represented by three elements, the plate, the jelly and the angioblasts.

The First, Umbilical, Circulation. Fig. 1

The angioblasts, including those of the heart, unite to form cords which hollow to give blood vessels, at first irregular and plexiform, but soon forming regular channels. An aorta passes caudally to become continuous with an umbilical artery formed by angioblasts in the connecting stalk aligned beside the allantois. Angioblasts are also formed in the chorion and chorionic villi, where they build up extensive vascular plexuses. An umbilical vein leads back from the stalk along the body wall of the embryo to the heart region from which the aorta begins as the primary loop. The vessel walls are of endothelium only, but this is contractile so that the plasma contents can at least be driven to and fro. There are no blood corpuscles at this stage, for there are, as yet, no connections with the vessels of the yolk sac.

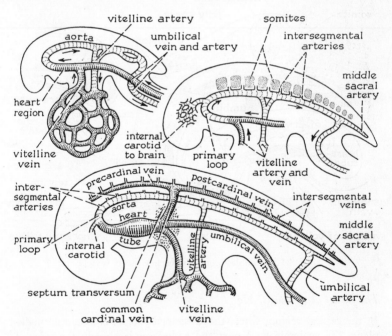

Fig. 4.2. Umbilical and vitelline circulation.

The Vitelline Circulation. Fig. 2

The network of vessels on the yolk sac is now joined to the umbilical circulation. A vitelline artery or plexus of vessels arises from the aorta and passes down the yolk stalk, while a vitelline vein passes back to the embryo to join the umbilical vein and form the vitello-umbilical trunk, which will eventually take part in the formation of the heart. The red corpuscles, originally formed in the blood islands, can now be washed into the general circulation.

Intersegmental Arteries. Fig. 2

The angioblasts spread throughout the jelly spaces of the embryo so as to provide a capillary circulation. Enlarged channels between the somites form the intersegmental arteries, whose arrangement is preserved to some extent in the intercostal and lumbar arteries of the adult. Cranially the aorta is prolonged as an internal carotid to supply the head structures, especially the rapidly developing brain. Caudally it is prolonged as a middle sacral artery, at this stage double, for the right and left arteries unite later.

Great Veins of the Embryo. Fig. 2

Corresponding to the intersegmental arteries, intersegmental veins develop in series between the somites. Blood from these is collected by pre- and post-cardinal (anterior and posterior cardinal) veins which join to form a common cardinal (duct of Cuvier) opening into the heart. Thus on each side three large veins meet, the umbilical from the connecting stalk, the vitelline from the yolk sac, and the common cardinal draining the embryonic tissues. These join in the septum transversum which serves as a bed leading them to the heart.

Embryonic Circulation. Fig. 2

Thus oxygenated blood reaching the heart by umbilical veins mixes with venous blood from the body of the embryo, led in by common cardinals and from the yolk sac by vitelline veins. It all leaves the heart by the aortae and is distributed to the body by intersegmental vessels and internal carotids, the yolk sac by vitelline arteries, and the chorion and chorionic villi by umbilical arteries.

Median Vessels. Fig. 3

The rudiments of the vascular system are bilateral, but in several regions the right and left elements soon fuse. Thus the right and left

vitello-umbilical trunks, partly united from the time of their first formation, fuse to form the endothelial heart tube. The right and left
aortae, at this stage consisting of simple endothelial walls only, unite
ventral to the notochord, between it and the gut, but the primary loops
remain separate. The two vitelline arteries unite to form a single vessel
springing from the single aorta. The two umbilical veins fuse in the

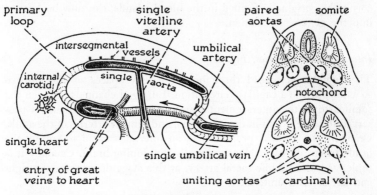

Fig. 4.3. Formation of median vessels.

connecting stalk, but their intraembryonic parts remain separate so that
the single vein divides as it enters the embryo.

Coelom. Fig. 4

As first formed the coelom is a simple, flat, U-shaped tube, open at
both ends into the extraembryonic coelom. But as the embryo folds off,
the cranial part of the coelom which crosses the midline becomes bent
ventrally, and eventually expanded round the developing heart tube to
form the pericardial cavity. Caudal to the pericardial cavity the coelom
is narrowed to form the pericardio-peritoneal canals which connect
the pericardial cavity with the abdominal part of the coelom. Thus a
passage enters the embryo from the extra-embryonic coelom of the one
side, passes through the abdominal part of the intraembryonic coelom
and pericardio-peritoneal canal of that side, crosses the midline by the
pericardial cavity, and leaves by the other side.

Septum Transversum and Great Veins. Fig. 4

The septum transversum is now a clearly demarcated mass of mesoderm. It lies between the pericardial cavity and heart cranially and the
abdominal coelom and yolk stalk caudally. Dorsally it reaches the fore-

gut and pericardio-peritoneal canals, while ventrally the amnion is attached to it. It receives the six great veins, the common cardinals winding round the canals, the umbilicals running in the body wall lateral to the abdominal coelom, and the vitelline veins running up the yolk stalk. The septum is, in fact, a mesodermal bed placed across the coelom to allow these vessels to unite in the midline to form the heart tube, whose caudal end is embedded in the septum.

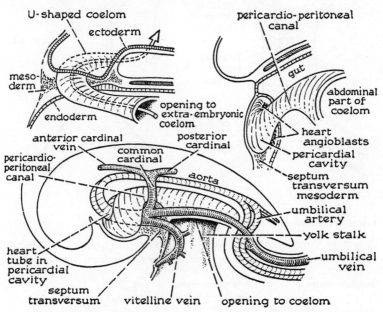

Fig. 4.4. Septum transversum and union of the great veins.

Y-Shaped Heart Tubes. Fig. 5

When the endothelial heart tubes first hollow out from the heart angioblasts they lie side by side. Cranially they soon unite, but caudally where they receive the great veins they are still widely divergent, so that the heart is shaped like an inverted Y. The veins enter the limbs of the Y and the primary loops of the aorta arise from its stem.

Primary Chambers of the Heart. Fig. 6

The Y-shaped tube is relatively narrow, but it has a thick covering of jelly and coelomic wall. Thus the heart as a whole appears as a thick Y projecting into the pericardial part of the coelom. It is demarcated by grooves into four primary chambers. The sino-auricular chamber is the most caudal, making the limbs of the Y and receiving the great veins. Bulges on either side, the right and left auricular swellings, indicate the future auricles. The two ventricles are arranged tandem fashion, the left caudal to the right, in the stem of the Y, and a short truncus leads from the right ventricle to the primary loops of the aorta.

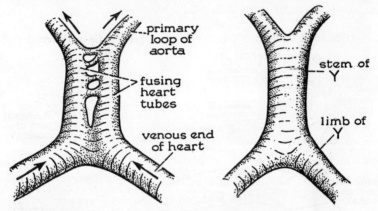

Fig. 4 5. Fusion of endothelial heart tubes.

Chest Wall and Heart Wall. Fig. 7

A section through the stem of the Y shows the uniting endothelial tubes surrounded ventrally and laterally by the coelom. The coelomic

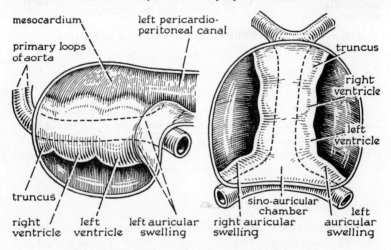

Fig. 4.6. The Y-shaped heart in its pericardial cavity.

wall next the ectoderm is thin, constituting with the ectoderm the early chest wall, so thin that the developing heart can be seen through it. The coelomic wall next the heart tubes, thickened earlier as the cardiogenic plate, is now differentiating as the outermost layer of the heart wall, the myoepicardial mantle, so called though it is still a single sheet.

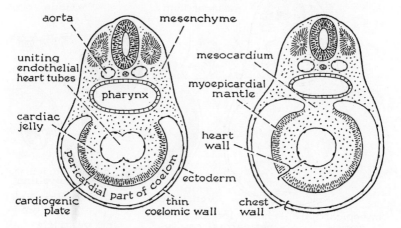

Fig. 4.7. Myoepicardial mantle and mesocardium.

Between the mantle and the endothelial tube the jelly space has been invaded by a few mesodermal cells to give a mesenchyme, the cardiac jelly, continuous with the rest of the embryonic mesenchyme.

The Mesocardium. Figs. 7 & 8

The heart is at first widely attached by mesenchyme to the pharynx. Later the coelom spreads round the dorsal surface of the heart from either side, narrowing the mesenchymal attachment to give a relatively thin mesocardium. Later still the coelomic extensions meet and fuse, cutting through the mesocardium and freeing the heart from the pharynx. The new opening is the transverse sinus of the pericardial cavity.

The Heart Beat. Fig. 8

The spread of the coelom between the heart and the pharynx carries the myoepicardial mantle towards the median plane. Here the approaching edges fuse to complete a tube of mantle tissue round the heart. Myofibrils develop in the cells and it begins to contract, irregularly at first, later with a steady beat. Meanwhile the cardiac jelly has become, by thickening of its ground substance, a resilient layer which keeps the heart lumen closed behind the blood as the muscle drives it along.

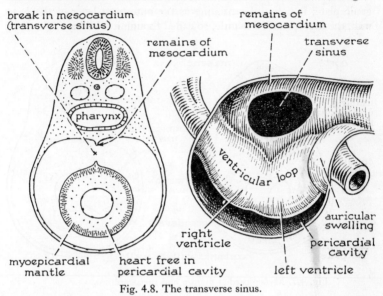

Fig. 4.8. The transverse sinus.

The Coelomic Fluid. Fig. 6

The heart lies in the pericardial cavity; the cavity communicates on either side, through the pericardio-peritoneal canals, with the abdominal coelom, and thus with the extra-embryonic coelom (fig. 3.5.). It has been suggested that the coelomic fluid, driven to and fro by the heart, carries oxygen and nutriment into the interior of the embryo. This may be a significant factor in nutrition of the embryo during the period lying between the closure of the neuropores cutting off the amniotic cavity from the neural canal and the establishment of a blood circulation.

The Free Heart. Fig. 8

Destruction of the mesocardium leaves the heart attached only at its two ends. Otherwise it hangs freely in the pericardial cavity and can now grow much faster than the rest of the embryo and become tortuous. The two ventricles elongate to give a ventricular loop, projecting forwards and to the right. This is the first important asymmetry of the embryo.

Rotation of the Left Ventricle. Fig. 9

The left ventricle, originally placed in the median plane, with the blood running cranially (fig. 6) is rotated so that the blood runs caudally and to the right. The opening from the sino-auricular chamber to the left ventricle *is* carried to the left, so that it comes to open from the left auricle. At the same time it is narrowed and elongated to an auriculo-ventricular canal. The two auricles, previously indicated by right and left swellings of the sino-auricular chamber, are now demarcated from the sinus venosus by indentations.

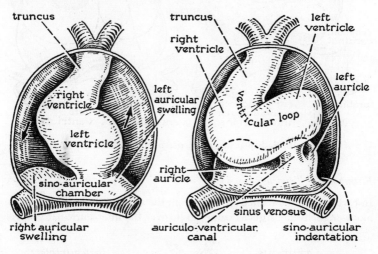

Fig. 4.9. Ventricular loop and separation of the auricles.

The S-Shaped Heart. Fig. 10

(atrium)

Whereas the indentation demarcating the right auricle from the sinus is relatively shallow that on the left is deep. Thus the sinus comes to open into the right and not the left auricle. The blood on reaching the sinus now passes successively through the right auricle, left auricle, left ventricle, right ventricle and truncus, following a twisted S-shaped course swinging first to the left and then to the right.

Rising of the Auricles. Figs. 9 & 10

At first the auricles, or auricular swellings, lie caudal to the ventricles (fig. 9), but later they pass dorsal to the ventricles and finally cranial to them (fig. 10). Both the sinus venosus and the truncus become elongated as the relative position of the auricles and ventricles change.

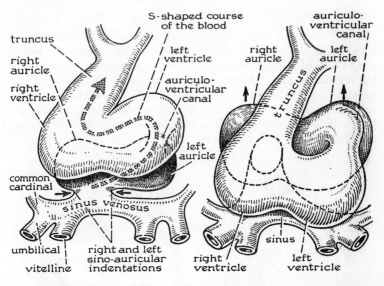

Fig. 4.10. S-shaped heart and rise of the auricles.

Fitting of the Truncus. Fig. 11

In early stages the truncus was far from the auricles, but as the auricles move cranially the truncus comes to lie over the inter-auricular groove.

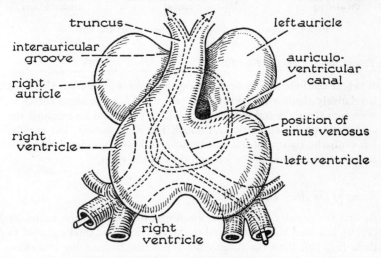

Fig. 4.11. Fitting the truncus into its groove.

Folding of the heart is completed by the truncus slipping into the groove, a position it retains permanently.

The Ventricular Walls. Fig. 12

The bulging parts of the ventricular walls become thickened primarily by an increase in their middle layer, the cardiac jelly. Branching in-growths from the outer layer, the myoepicardial mantle, spread into the jelly. When they approach the endothelial lining this inner layer responds by growing out amongst the in-growths so as to clothe each one with endothelium. Soon after, the jelly disappears leaving the ventricular wall a spongy mass of contractile tissue derived from the mantle, with endothelium lined spaces between the strands. Thus the contractile cells are all near the blood circulating through the heart lumen and are efficiently nourished. The interior of auricles becomes roughened by a similar but less marked process. The other parts of the heart remain smooth.

Interventricular Foramen. Fig. 12

In passing from the left to the right ventricle the blood has to pass through the relatively narrow interventricular foramen. This region is

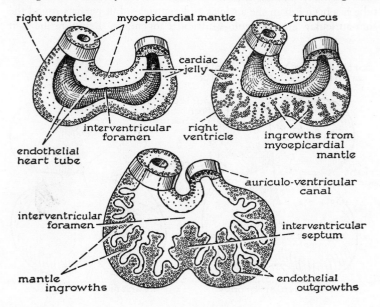

Fig. 4.12. Spongy walls of the ventricles.

not involved in the spongework formation of the ventricles. A solid, thick, eventually muscular, interventricular septum is built up between the expanding ventricles, and the blood passes across its free margin.

Squaring Off The Heart. Fig. 13

The heart is now ready to divide into right and left sides, using the grooves between the right and left auricles and the interventricular septum as a start. The auriculo-ventricular channel shifts to the mid-plane, becoming short, wide and straight as it moves. In this position it lies directly over the interventricular septum, opening from both auricles into both ventricles. A deep coronary groove separates the auricles from the ventricles and the heart has a symmetrical, squared off appearance.

Interior of the Heart. Fig. 14

Excision of the ventral part of the heart opens the interior. The ventricular wall is thick and spongy, the auricular wall thin, but of similar structure. The sinus venosus still opens into the right auricle, but the auriculo-ventricular passage lies in the median plane. The cardiac jelly that originally padded the whole length of the interior of the heart tube has been lost in the auricles and ventricles, but persists round the sino-auricular and auriculo-ventricular openings. Being resilient it

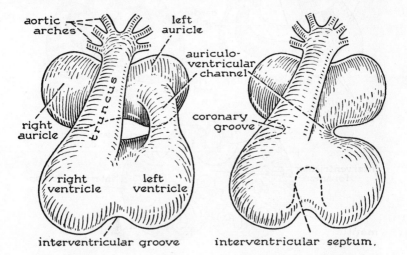

Fig. 4.13. The auriculo-ventricular channel moved to the mid-plane.

keeps the openings closed except when blood is being forced through them, so directing the blood flow.

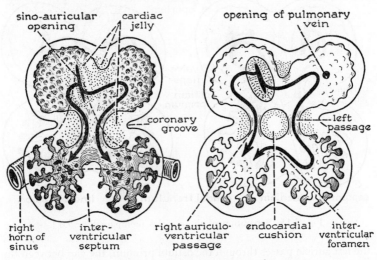

Fig. 4.14. Division of the auriculo-ventricular passage.

Endocardial Cushions. *Fig. 14*

A pair of soft elevations, the endocardial cushions, grow into the auriculoventricular passage from its anterior and posterior walls. The cushions, which eventually fuse to form the septum intermedium, subdivide the passage into right and left passages, the first definite indication of double circulation in the heart. The interventricular foramen now lies between the interventricular septum and the fusing cushions, and only half the blood, that coming from the left ventricle, has to pass through it.

The Pulmonary Vein. *Fig. 14*

A venous plexus develops round the expanding lung buds and opens into the left auricle by a single opening. The amount of blood delivered by this vein is relatively small, most of the blood in the auricle coming from the sino-auricular opening.

The Septum Primum. *Fig. 15*

A little cardiac jelly is found between the right and left auricles, opposite the interauricular groove. This is drawn out into a thin sheet, the septum primum, with a thickened edge. Blood going from the right to

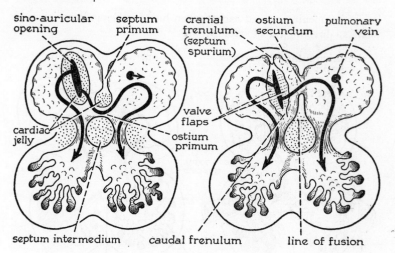

Fig. 4.15. Separation of the auricles (compare Fig. 9.8.).

the left auricle passes through the ostium primum, the gap between the edge and the septum intermedium.

The Ostium Secundum. Fig. 15

The septum primum fuses with the septum intermedium, so obliterating the ostium. This would cut off the blood from the left auricle (apart from a trickle from the pulmonary vein), if no other provision were made. But a new opening, the ostium secundum, is cut through the septum primum, allowing the blood to pass. This is essential as each chamber of the heart must be kept filled with circulating blood if it is to grow properly.

The First Valve. Fig. 15

The sino-auricular opening has, from the time of its first formation, been guarded by lips supported by cardiac jelly. These lips thin and toughen to make a pair of flaps which together act as a sino-auricular valve. The ends of the valve are drawn out into frenula, the upper frenulum, or septum spurium, being especially prominent.

The story of the heart is continued in Chapter 14, p. 134

5

The Pharynx

So far the head has been a smooth prominence overlying the brain, but now it becomes modified by the formation of the paired sense organs and the branchial arches. The optic diverticulum arises as a hollow outgrowth from the optic groove of the forebrain. Later a

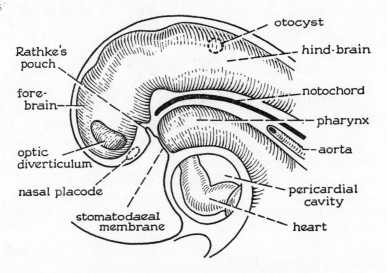

Fig. 5.1. Optic diverticulum, otic and nasal placodes.

thickening of the ectoderm, the otic placode, overlying the hindbrain, indicates the acoustico-vestibular apparatus. A similar thickening, the nasal placode, overlies the forward end of the forebrain near Rathke's pouch.

Optic Cup. Fig. 2

The end of the optic diverticulum expands to form a vesicle attached by a hollow stalk to the brain wall. The face of the vesicle sinks in to form the optic cup with an inner invaginated layer and an outer layer. At the same time a groove, the choroidal fissure, is developed on the ventral surface of the cup and stalk for the accommodation of an artery to the interior of the cup.

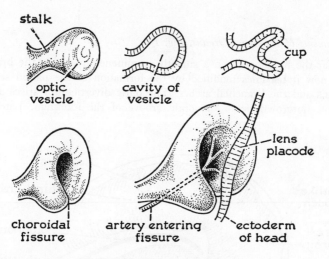

Fig. 5.2. Optic cup and lens.

Lens. Fig. 2

The ectoderm over the cup thickens to form a lens placode. The main rudiments of the eye are now laid down, the two layers of the cup, the stalk, the lens placode, the overlying ectoderm and the mesenchyme that surrounds and fills the cup.

Embryonic Hormones. Fig. 3

The growth of the eye is better understood than most embryonic processes as it is open to experiment. If an optic cup is cut out from one frog embryo of suitable age and inserted beneath the ectoderm of say, the tail of another, its presence induces the formation of a lens from the ectoderm. Conversely if a piece of tail ectoderm is exchanged for head ectoderm which, if it had been left in place, would have formed a lens, the tail ectoderm now forms a lens while the head ectoderm

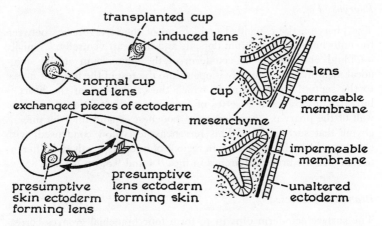

Fig. 5.3. Experiments on the optic cup.

forms ordinary tail skin. A cellophane sheet which allows medium-sized molecules to pass through it placed between the cup and the ectoderm does not interfere with lens formation, but a metal-foil which is impervious does so interfere. So it is believed that the optic cup secretes a hormone which acts on the overlying ectoderm, inducing lens formation, and it is probable that most embryonic growth processes depend on such hormones acting locally on tissue at the right stage of development.

Otocyst. Fig. 4

The earliest indication of the auditory apparatus is an ectodermal thickening, the otic placode, placed just caudal to the widest part of the hind-brain. The placode forms a pit and is finally cut off as a hollow otocyst.

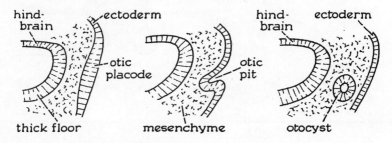

Fig. 5.4. Otic placode and otocyst.

Pharynx. Fig. 5

The pharynx, the expanded cranial end of the fore-gut, lies between the notochord and hindbrain dorsally and the heart ventrally. Cranially its blind end meets the ectoderm of the stomodeum at the stomodeal membrane. Caudally it opens into the rest of the fore-gut dorsal to the septum transversum in which the caudal end of the heart is buried. On either side a series of four outgrowths of the endodermal wall of the pharynx, the pharyngeal pouches, expand into the mesenchyme that surrounds it. This mesenchyme is now condensed, with closely packed cells, as is usual in regions of intense growth and differentiation (see the branchial arches in figs. 9.4 and 9.5, pp. 83 and 85).

Branchial Grooves. Fig. 5

The surface ectoderm dips in to form four branchial grooves corresponding to the four pharyngeal pouches of the underlying endoderm. The ectodermal groove and endodermal pouch meet, pinching out the intervening mesenchyme and forming a series of branchial membranes. In fishes these membranes eventually break down, putting the pharynx into communication with the exterior by a series of branchial clefts on each side, but in man the membranes normally never rupture, and the fourth pouch and groove never actually meet. Towards the stomodeum the pharynx is wide so that the oral membrane is much larger

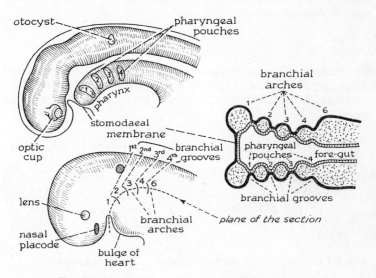

Fig. 5.5. Pharyngeal pouches and branchial arches.

than the branchial membranes. This membrane breaks down so that the mouth opens into the pharynx.

Branchial Arches. Fig. 5

Thus the pharyngeal region is set off into a series of branchial arches lying between the developing brain dorsally and the heart ventrally. Each arch consists of a core of mesenchyme, as yet undifferentiated, surrounded by a coat of epithelium partly ectodermal and partly endodermal in origin. The arches, which diminish in size craniocaudally, the first being much the largest, are numbered 1st, 2nd, 3rd and 4th. In some animals and occasional human embryos there is a 5th, but in most human embryos it never appears. Although it is not defined caudally from the body wall, the ridge behind the last groove is considered a 6th arch, since in later development it behaves as such.

Floor of Pharynx. Fig. 6

The ventral ends of the branchial arches turn into the floor of the pharynx. Opposite the first branchial pouches, a median thyroid diverticulum arises as an outpouching of the endodermal wall of the pharynx, and at its posterior end, where the pharynx joins the rest of the foregut, the respiratory diverticulum arises from a groove in its floor.

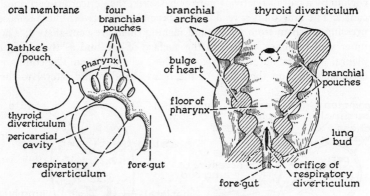

Fig. 5.6. Pharyngeal floor and its diverticula.

Separation of Trachea. Figs. 6 & 7

The respiratory diverticulum is narrow from side to side and opens into the pharynx by an elongated orifice. Right and left lung buds grow from the caudal end of the diverticulum. A groove developing on either

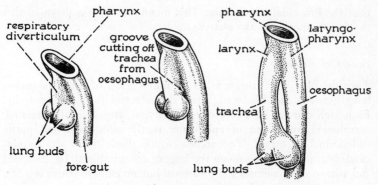

Fig. 5.7. Lung buds and trachea.

side meets to cut off a part of the diverticulum, which becomes the trachea, from the oesophagus, and both these tubes become elongated as the embryo grows. The most cranial part of the original diverticulum remains as the larynx, opening into the laryngo-pharynx.

Pericardio-peritoneal Canals. Fig. 8

It will be remembered that the pericardial part of the coelom surrounding the heart is connected to the main abdominal part of the coelom by a pair of relatively narrow passages, the pericardio-peritoneal canals. The lung buds spread in the mesenchyme between these canals, pressing on them from within and making them crescent-shaped. The inner wall becomes applied to the surface of the bud as the visceral pleura, while the outer wall becomes the parietal pleura. The common cardinal veins, on their way to the heart, wind round lateral to the

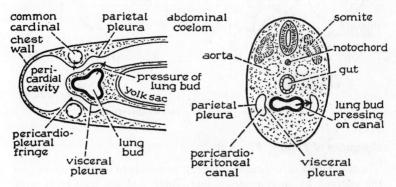

Fig. 5.8. Lung buds and pericardio-peritoneal canals.

canals, compressing them from without. A small ridge of tissue, the pericardio-pleural fringe, forms over the vein, and further narrows the passage between the pleural and pericardial cavities.

Aortic Arches. Fig. 9

The heart lies ventral to the pharynx and when the branchial arches have been formed the primary loop of the aorta is found curving dorsally beside the stomodeum in the mesenchyme of the first arch. A series of new vessels, the 2nd, 3rd, 4th and 6th aortic arches, develop in succession in the branchial arches from which they are named, the

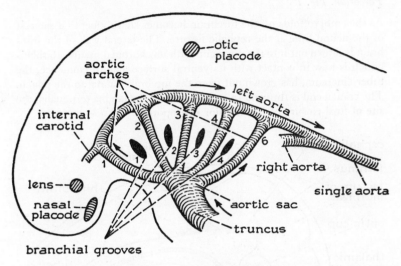

Fig. 5.9. Arterial arches, drawn as if all were present at one time.

primary loop itself forming the 1st. Not all are present at any one time, for the first two have degenerated before the last has appeared. Ventrally they spring from the aortic sac which caps the truncus and represents, in the human embryo, the ventral aorta of other animals such as the dogfish. Dorsally they collect in the right and left aortae which join behind the branchial region to form the single aorta. It is clear that the segmentation of the branchial mesenchyme into arches prepares it for the orderly development of these vessels and other branchial formations, nervous, skeletal and muscular, that will appear later.

The later changes in the pharynx and branchial arches will be found in chapters X, p. 103, Face and Nose; XI, p. 110, Mouth; XII, p. 118, Branchial Arches and XIII, p. 125, Pharyngeal Derivatives.

6

The Brain and Spinal Nerves

Forebrain. Fig. 1

As the embryo folds off the forebrain becomes sharply bent on the rest of the neural tube at the cephalic flexure. The lateral wall of the forebrain is drawn out into the optic cup with its choroidal fissure. Rathke's pouch is now in contact with its ventral surface and an eminence, the tuber cinereum, has grown out from the forebrain caudal to the pouch. The cranial end is thin and so recognisable as the lamina terminalis, the site of final closure of the anterior neuropore.

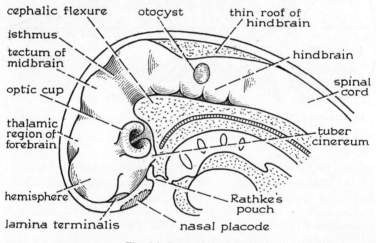

Fig. 6.1. Parts of the brain.

Cerebral Hemisphere. Figs. 1 & 2

There grow out from the forepart of the forebrain a pair of hemispheres, which are at this stage truly hemispherical. The hollow in each hemisphere is the lateral ventricle, which communicates with the main cavity of the forebrain, the third ventricle, by the interventricular foramen. The hemisphere grows from only the dorso-cranial part of the forebrain, leaving caudal to it a thalamic portion from which the thalamus will develop later.

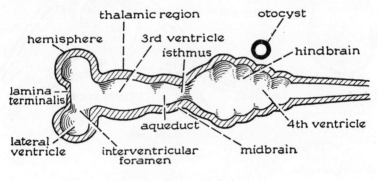

Fig. 6.2. Cavities of the brain.

Midbrain. Figs. 1 & 2

The midbrain forms a curving tube of much greater extent dorsally than ventrally where it is sharply bent at the cephalic flexure. Its lumen, still wide at this stage, is the cerebral aqueduct. Its roof, the tectum, is smooth, for the colliculi have not yet appeared. Caudally it is sharply defined from the hindbrain by a constriction, the isthmus.

Hindbrain. Figs. 1 & 2

The hindbrain is widened to a diamond shape, with the widest part about 1/3 of the length from the isthmus, and the otocyst lies caudal to the widest part. The roof is stretched to form a thin membrane, through which the cavity, the fourth ventricle, can be seen. The floor is thrown into a series of waves, better marked on the inner than the outer surface, which later disappear and whose significance is unknown. Caudally the hindbrain passes into the spinal cord without any sharp boundary.

Neural Crest. Fig. 3

Soon after the neural folds have joined to form the neural tube, cells migrate out from the tube and condense to give two bands of tissue, the neural crests. These lie between the dorsal parts of the tube and the somites on either side, extending cranially to near the isthmus.

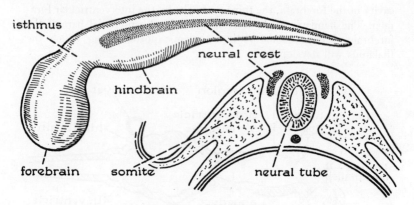

Fig. 6.3. Early neural crest.

Sensory Ganglia. *Fig. 4*

It has been seen that there are two great segmentation systems in the embryo, an earlier somite system in the trunk and a later branchial arch system in the head, and both affect the neural crest. In the trunk the crest breaks up into blocks which eventually form the sensory spinal ganglia. In the head four blocks are found, each at the dorsal end

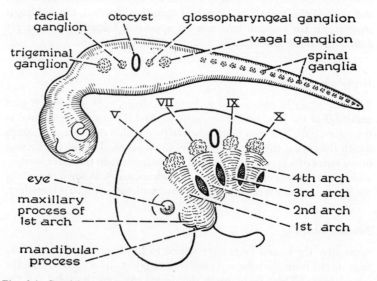

Fig. 6.4. Cranial and spinal sensory ganglia. The roman figures indicate the cranial ganglia.

of a branchial arch. These are the trigeminal, facial, glossopharyngeal and vagal ganglia, corresponding to arches 1-4. The first two lie cranial to the otocyst, the other two caudal to it. The facial ganglion comprises three elements, facial, acoustic and vestibular, which separate later.

Spinal Cord. *Fig. 5*

The spinal cord now consists of thicker side walls joined by thinner roof and floor plates. Each side wall is subdivided by the limiting sulcus (sulcus limitans) into a dorsal alar lamina and a ventral basal lamina, both running the length of the cord. The alar lamina with the neural crest is concerned with building the sensory and associative apparatus and the basal lamina with the motor apparatus of the nervous system.

Ependymal Layer. *Fig. 5*

The innermost cells next to the central canal become arranged as a simple columnar ciliated epithelium, the ependymal layer. These cells divide rapidly, the daughter cells passing out into the substance of the wall to form the middle or mantle layer.

Mantle Layer. *Fig. 5*

Some of the cells of the mantle layer remain, for the time, undifferentiated, but the others differentiate into the early neurons (neuroblasts) and supporting cells. The supporting cells (spongioblasts) develop processes which penetrate the whole thickness of the wall in both directions and spread on its surfaces to form the internal and external limiting membranes. The internal membrane is pierced by the cilia of the ependymal layer. The neurons develop axons which enter the outer, marginal, layer.

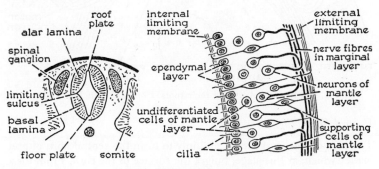

Fig. 6.5. Wall of the spinal cord.

Marginal Layer. Fig. 5

The marginal layer contains no cell bodies, being made up entirely of the processes of neurons and supporting cells. The three main layers of the cord are now established, the ependymal which persists as such throughout life, the mantle which becomes the grey matter and the marginal which becomes the white matter of the adult. At this stage the structure of the brain wall and its diverticula is similar to that of the cord.

Spinal Nerves: Dorsal Roots. Fig. 6

The cells of the spinal ganglia differentiate like those of the mantle layer into supporting cells and nerve cells. Each nerve cell develops a process from either end, a dendrite growing peripherally to bring in impulses from the body wall, and an axon growing centrally to enter the substance of the neural tube and end in the mantle layer of the alar lamina. All the neurons of the spinal ganglion are thus, at this stage, bipolar. The processes together form the dorsal root of a spinal nerve, and where they enter the cord they stimulate the mantle layer to expand into a dorsal horn of grey matter.

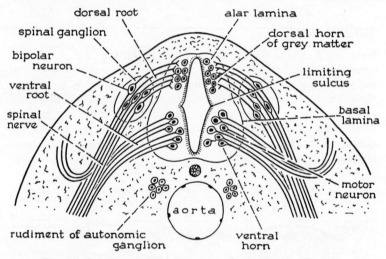

Fig. 6.6. Spinal nerves and their roots. Compare with Fig. 9.13.

Ventral Roots. Fig. 6

The cells of the basal lamina multiply to form the ventral horn of grey matter which, at this stage, bulges the surface of the cord. From these cells axons grow out to each somite forming the ventral roots of the spinal nerves. A few cells from the spinal ganglia, originally of the neural crest, migrate ventrally to form the rudiments of the autonomic ganglia, but these are not yet differentiated.

Cranial Nerves. Fig. 7

The olfactory placode is now sinking below the general surface of the ectoderm to form the olfactory pit, but the olfactory nerve has not yet appeared. The place of the optic nerve is indicated by the optic stalk, but there are as yet no fibres in it. The other cranial nerves grow as do the spinal, sensory fibres springing from ganglion cells and motor fibres, from the mantle layer of the brain stem. The trigeminal ganglion is the largest of the series and gives its three main branches, ophthalmic, maxillary and mandibular, the first passing dorsal to the eye and the others into the maxillary and mandibular processes into which the first branchial arch has now divided. The facial supplies the second arch, its ganglion, the geniculate, being concerned with taste. The acoustico-vestibular ganglion has now become differentiated from the facial and

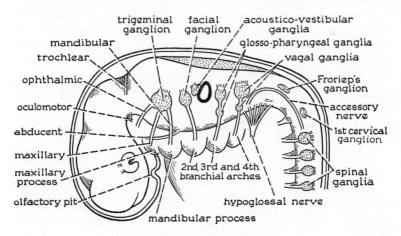

Fig. 6.7. Cranial and spinal nerves.

is closely applied to the otocyst. Behind the otocyst the glossopharyngeal supplies the third arch, its ganglia split into an upper for general sensation and a lower for taste. The vagus supplies the remaining arches, and its ganglion is similarly split. The accessory nerve arises from the cervical part of the spinal cord and curves cranially beside the caudal part of the hindbrain to approach the vagal roots.

Motor Nerves and Transitory Ganglia. *Fig. 7*

The oculomotor nerve grows from the midbrain in the cephalic flexure, the trochlear from the roof of the hindbrain just behind the isthmus, and the abducent from the floor of the hindbrain. A series of roots from the hindbrain in line with the oculomotor and abducent represents the hypoglossal. Two transitory ganglia, the ganglion of Froriep at the level of the hypoglossal and the spinal ganglion of the first cervical nerve, are distinguished at this stage but are lost later.

Experiments on Spinal Nerves. *Fig. 8*

That the neural crest material does in fact form the spinal ganglia, sensory nerves and autonomic system can be demonstrated by its removal. A cut removing the dorsal part of the body of an embryo removes the crest completely and none of these structures develop. Substitution of a series of small somites from the caudal end of one embryo for some of the larger somites of the cranial end of another embryo leads to the development of additional spinal nerves. This shows that the somites impose their segmentation on the nervous system and not vice versa, confirming the observation that the paraxial mesoderm segments before the spinal nerves appear. That the nerve fibres do, in fact, grow out freely into the mesenchyme is demonstrated by the culture of small fragments of spinal cord in nutrient media. Processes headed by amoeboid masses of protoplasm are seen to sprout from the fragments, each amoeboid mass appearing to feel its way forward, laying down a fibre behind it.

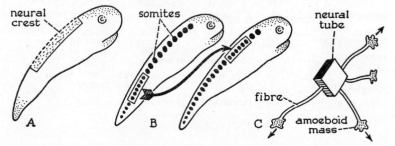

Fig. 6.8. Experiments on nerves: (a) Removal of neural crest. (b) Transplantation of somites. (c) Growth of fibres.

Direction of Growth. *Fig. 9*

If a circular wire loop is dipped into a protein-containing nutrient mixture it picks up a film of the mixture. A fragment of spinal cord can be dropped into the film and the film coagulated by the addition of a little embryo extract. Nerve fibres are found to grow out into the coagulum in all directions. But if the loop is made triangular the fibres grow out mainly in three directions, towards the nearest parts of the wire. Probably the growth of the fragment causes desiccation and contraction of the film, setting up internal strains and inducing rearrangement of its macro-molecules. The nerve fibres are presumably guided in certain definite directions by the molecular arrangement.

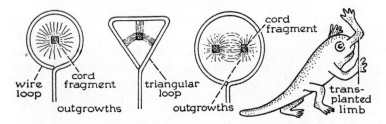

Fig. 6.9. Experiments on direction of nerve growth.

Peripheral Nerves. Fig. 9

If two cord fragments are planted in the same film the fibres grow towards each other forming a pattern like the lines of force between the poles of a magnet. Here again the fibres are following pathways determined by the submicroscopic arrangement of the material in which they are growing. It seems likely that in the living embryo the submicroscopic arrangement of the ground substance and the resistances set up by developing organs guide the nerves on their way. If a developing limb, before any nerves have reached it, is transplanted to the snout of a host embryo it continues its development and is later found to contain median, radial and other nerves, normal in position but derived from the nerves of the head. Even the number of fibres in each nerve may be normal, though derived by branching from a few cranial fibres. This suggests that the nerves are controlled in development by the structures amongst which they grow.

7

The Liver and Stomach

Hepatic Diverticulum. Fig. 1

It has been seen that the septum transversum is a mass of mesenchyme placed between the pericardial cavity and the yolk stalk. The most caudal chamber of the heart, the sinus venosus, is embedded in it and the six great veins use it to reach the heart. A proliferation of the mesenchyme is the first indication of the future liver, soon followed by an endodermal outgrowth, the hepatic diverticulum, from the yolk stalk. The diverticulum burrows into the septum between the vitelline veins, which are still plexiform at this stage. The gall bladder appears at the same time in continuity with the main hepatic rudiment.

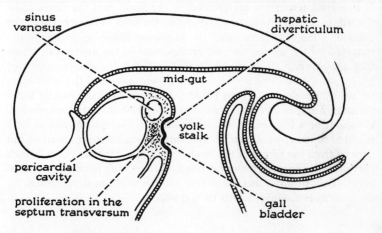

Fig. 7.1. Rudiments of the liver and gall bladder.

Hepatic Trabeculae. Fig. 2

Epithelial trabeculae spread out from the hepatic diverticulum into the greater part of the septum. They reach the vitelline veins whose endothelial walls respond by clothing the trabeculae, so that the epithelial cells are enmeshed in a venous network.

Spread of Liver. Figs. 2 & 3

The liver grows rapidly, coming to form right and left lobes that bulge into the main coelomic cavity and a pair of dorsal lobes that project

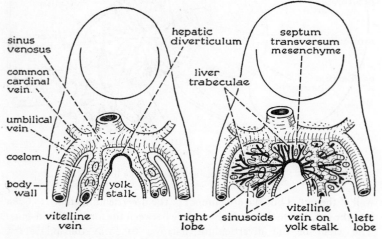

Fig. 7.2. The liver trabeculae.

into the pericardio-peritoneal canals beside the lung buds. The part of the septum next the pericardial cavity is not invaded.

Dilated Veins. Fig. 3

The main channels of the vitelline veins are pressed dorsally by the growth of the liver and come to lie, as the hepato-cardiac veins, in the dorsal lobes. Here they are dilated and are well placed for exchange between the blood they contain and the coelomic fluid in the pleuro-pericardial canals. The umbilical veins lie in the lateral body wall in a very loose mesenchyme, Wharton's jelly. They also are dilated, and here again there may be exchange of substances between their blood and the coelomic fluid.

Transfer of Diverticulum. Fig. 4

When first formed the hepatic diverticulum springs from the yolk stalk, but as folding off is completed much of the stalk is taken into the

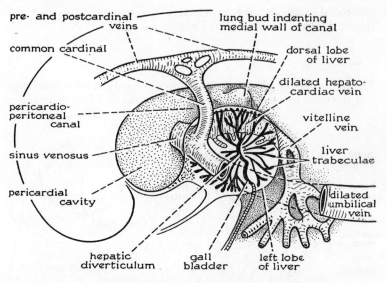

Fig. 7.3. The major veins and the liver.

wall of the gut. Thus the attachment of the diverticulum comes to open into the gut, and to define the boundary between the fore- and mid-guts. The stem of the diverticulum narrows to form the bile duct. As the liver enlarges the yolk sac dwindles and the liver takes over the formation of blood cells, a function it retains until the bone marrow takes over in turn.

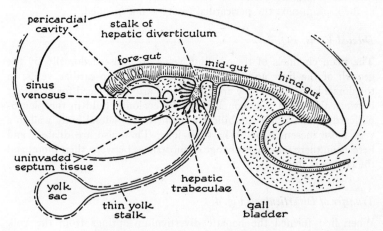

Fig. 7.4. Diverticulum transferred to gut.

Attachment of Gut. Fig. 5

So far the gut has been short and has followed the curve of the embryo, lying ventral to the neural tube, notochord and aorta. It is in fact attached to the aorta by a broad mass of mesenchyme, but on either side it is separated from the body wall by the coelom.

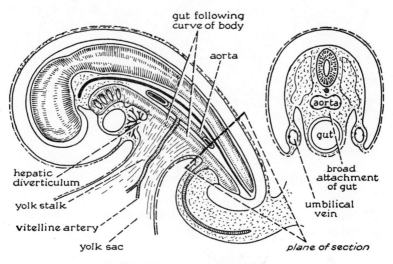

Fig. 7.5. Gut attached to aorta.

Primary Loop of Gut and Mesentery. Fig. 6

Now the gut begins to grow faster than the rest of the embryo, but having no function at this stage, remains narrow. Thus it comes to form a primary loop projecting far into the extraembryonic coelom. The attachment to the aorta is stretched out to a mesentery, which at this stage extends all the way from the pharynx to the cloaca. The vitelline arteries fuse to form a single vessel, the superior mesenteric, which lies in the axis of the loop. A caecal diverticulum grows from the returning limb of the loop. New, unpaired arteries, the coeliac and inferior mesenteric, later supply the fore- and hind-guts. On either side of the mesentery the intra- and extraembryonic coeloms are still widely continuous.

Stomach. Fig. 7

The fore-gut as it passes dorsal to the liver enlarges to form the spindle-shaped stomach, not yet sharply defined from the duodenum. For a

D

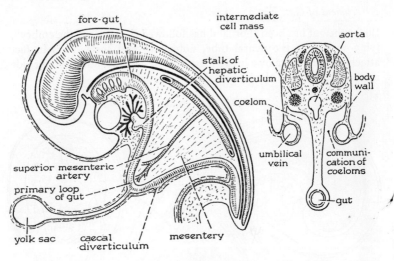

Fig. 7.6. Formation of the mesentery (compare Fig. 9.13.).

time the developing left lung bud is in direct contact with the oeso-phageal end of the stomach.

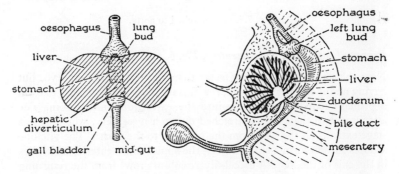

Fig. 7.7. Stomach and bile duct.

The Lesser Sac. Fig. 8

The fore-gut, together with the stomach and the lung buds, is clothed by a thick layer of mesenchyme, continuous with the septum trans-versum and mesentery. A depression, the lesser sac, develops in the mesenchyme covering the right surface of the stomach. It burrows cranially so that its tip comes to lie between the oesophagus and the

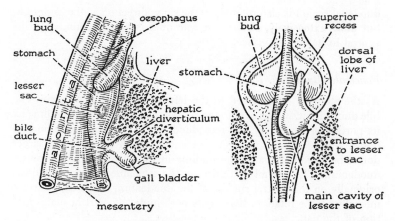

Fig. 7.8. The lesser sac, in right and dorsal view.

right lung bud. From the original pocket, which forms the superior recess of the lesser sac, a new diverticulum spreads dorsal to the stomach to form the main cavity of the sac.

Rotation of the Stomach and Duodenum. Fig. 9

The stomach bulges to the left and rotates through a right angle so that what was its ventral border comes to look to the right as the lesser curvature and its originally dorsal border to the left as the greater curvature ('the stomach falls on to its right side'). This rotation helps to carry the lesser sac to the left and to expand its main cavity. The

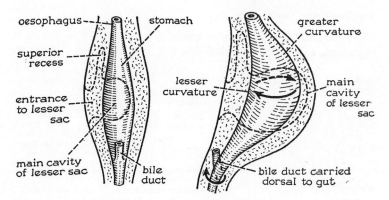

Fig. 7.9. Effects of Rotation, ventral views.

duodenal region is also rotated, in the same direction as the stomach, but more extensively, so that the bile duct is carried round from its ventral to its dorsal surface.

Pancreatic Buds. Fig. 10

As the stomach is displaced to the left, the duodenum, carrying the bile duct with it, moves over to the right, still attached to the midline by its portion of mesentery, the mesoduodenum. Two pancreatic buds, a dorsal and a ventral, arise from the endoderm of the gut. The larger dorsal bud grows towards the midline in the substance of the mesoduodenum. The ventral bud arises at the attachment of the bile duct, originally on the ventral surface of the duodenum, but is later carried medially by the rotation. Eventually both buds take part in the formation of the pancreas.

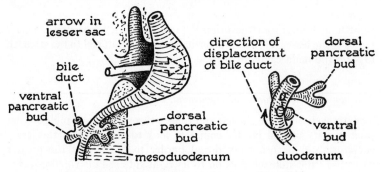

Fig. 7.10. The pancreatic buds (compare Fig. 9.11.).

Vitelline Cross Anastomoses. Fig. 11

The right and left vitelline veins were displaced by the liver during its expansion, and the parts included in the liver expanded to make the

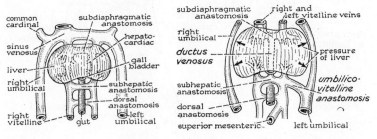

Fig. 7.11. The vitelline veins.

hepato-cardiac veins, but they still drain blood from the yolk sac and intestine towards the heart. Now a series of cross anastomoses forms between the veins. The most caudal passes dorsal to the gut, the middle, the subhepatic, ventral to the gut near the liver, and the most cranial, the subdiaphragmatic, again ventral to the gut, in the substance of the liver near the sinus venosus.

Ductus Venosus. Figs. 11 & 12

A new channel, the ductus venosus, forms in the median plane within the liver substance (see fig. 9.10, p.91). The rapidly growing liver presses on the umbilical veins. The right umbilical disappears, leaving the left to bring all the blood back from the connecting stalk to the heart. An umbilico-vitelline anastomosis forms between this vein and the left vitelline at the level of the subhepatic anastomosis. The venous pathways connecting the left umbilical vein with the cranial end of the right vitelline are enlarged and smoothed out. The left umbilical cranial

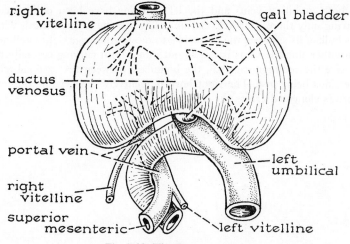

Fig. 7.12. The ductus venosus.

to the umbilico-vitelline anastomosis is obliterated so that now all the oxygenated blood from the connecting stalk crosses the midline and reaches the heart by the ductus venosus and right vitelline vein.

Sinus Venosus. Fig. 13

The sinus venosus originally had six veins opening into it, the paired vitellines, umbilicals and common cardinals. But now the terminal parts

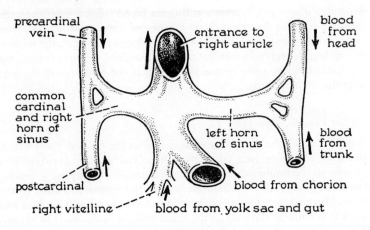

Fig. 7.13. Veins to heart.

of both umbilical veins and the left vitelline have been obliterated, leaving only three. Further, the sinus has become asymmetrical, with the bulk of its space on the right side, to receive the large right vitelline. On either side the sinus is drawn out into a horn curving cranially to receive the common cardinal, the boundary between the cardinal and the sinus being lost. The right horn is relatively short, while the left horn is elongated across the midline.

8

Mesonephros and Limb Buds

Intermediate Cell Mass (Intermediate Mesoderm). Fig. 1

The intermediate cell masses separate from the somites and lateral plate mesoderm to form a segmented series of mesodermal blocks lying dorsolateral to the paired aortae.

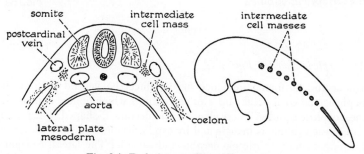

Fig. 8.1. Early intermediate cell masses.

Mesonephric Duct. Fig. 2

The most cranial masses, those of the upper cervical region, soon disappear. No clearly defined structure comparable to the pronephros of lower vertebrates develops in the human embryo, but the duct of the mesonephros probably begins as a pronephric duct. The masses of the lower cervical and upper thoracic regions fuse to form a continuous nephrogenic cord. This cord splits into dorsal and ventral parts. The dorsal part hollows to form the mesonephric duct. Once the duct is formed its free end grows rapidly through the mesenchyme, just beneath the ectoderm, towards the cloaca (it can be stained in the living embryo and watched as it makes its way). On reaching the cloaca the duct turns ventrally and eventually opens into it. Thus the cloaca now has four openings into it, the hind-gut, the allantois and on either side, the mesonephric ducts. It is, however, still cut off from the exterior by the intact cloacal membrane.

71

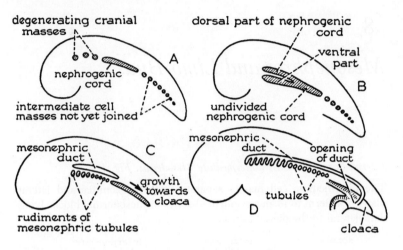

Fig. 8.2. Mesonephric duct and tubules.

Mesonephric Tubules. Figs. 2 & 3

Under the hormonal influence of the duct the ventral part of the nephrogenic cord breaks up into cell clumps which hollow to form the mesonephric tubules. These elongate and become S-shaped. The medial end of each tubule is invaginated by capillaries derived from the aorta, forming a mesonephric glomerulus and capsule. The lateral end joins the mesonephric duct. New tubules are added to those first formed by the nephrogenic tissue of the lower thoracic and upper lumbar segments, each tubule joining the mesonephric duct as this passes to the cloaca. A number of tubules, each with its own glomerulus and separate opening into the mesonephric duct, are formed in each segment.

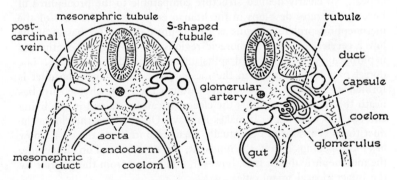

Fig. 8.3. The mesonephric components.

Mesonephros. Fig. 4

The tubules, glomeruli and duct of each side, together with the mesenchyme associated with them, constitute the mesonephros. This is, next the liver, the largest organ of the abdomen at this stage, projecting

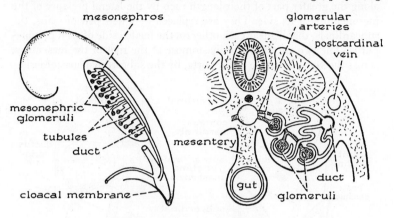

Fig 8.4. The mesonephros.

into the coelom, as a ridge beside the mesentery. The mesonephric duct, originally lying near the ectoderm, sinks towards the coelom and becomes included in the ridge. The postcardinal vein, also originally a superficial structure, lies in the root of the ridge, so closely applied to the mesonephros that the tubules indent its lumen.

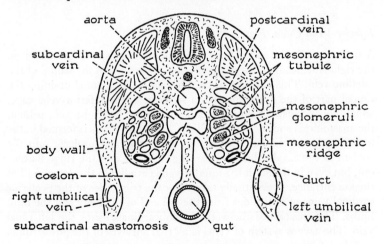

Fig. 8.5. The mesonephric ridges (compare Fig. 9.13.).

Subcardinal Veins. Fig. 6

The anterior (pre-) cardinal veins receiving blood from the head persist as the main channels draining the region cranial to the heart. But the posterior (post-) cardinal veins are compressed and finally obliterated along the greater part of their length each by the lateral pressure of the mesonephros at its side. They are replaced by a new pair of veins, the subcardinals, that lie near each other on the medial side of each mesonephros. The two subcardinals anastomose at the root of the mesentery, across the ventral surface of the aorta, by the subcardinal anastomosis.

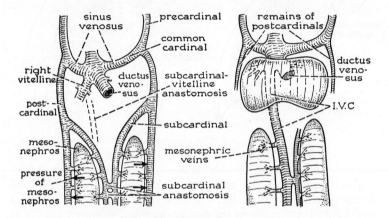

Fig. 8.6. Subcardinal veins and inferior vena cava.

Inferior Vena Cava. Figs. 6 & 7

A new channel, the subcardinal-vitelline anastomosis, is laid down on the right side, leading directly from the subcardinal to the stump of the vitelline vein. This enlarges to become the major channel draining the caudal part of the body, forming a segment of the inferior vena cava, which now consists of three parts derived from the right subcardinal, the anastomosis and the right vitelline. The anastomosis is formed in the mesenchyme of the entrance to the lesser sac, and projects into the coelom as the caval ridge (compare fig. 9.11.). Later the ridge flattens out, but even in the adult the inferior vena cava lies in the dorsal wall of the sac entrance. It eventually takes over the tributaries of the posterior cardinal veins of both sides from the tail end of the body and the hind limbs. The anastomosis between the subcardinals forms the left renal vein. The azygos system of veins is all that survives of the postcardinal vessels.

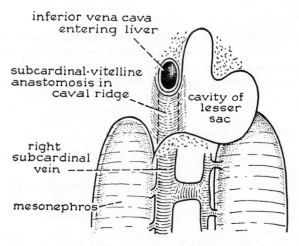

Fig. 8.7. The caval ridge and lesser sac.

Transference of Umbilical Artery. Fig. 8

When the mesonephric duct sinks deeply, away from the ectoderm, it comes into contact with the umbilical arteries as they curve ventrally beside the cloaca. New arteries are formed lateral to the ducts, while the old channels disappear, allowing the ducts to sink further towards the cloaca. Thus the umbilical arteries come to run lateral instead of medial to the ducts. This method of changing the course of an artery by replacing an old channel with one in a more convenient place is common in development.

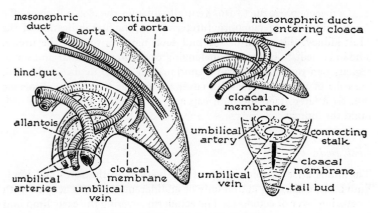

Fig. 8.8. Cloaca and cloacal membrane.

Tail Bud. Fig. 8

The tail bud projects as a conical prominence and in its interior extensions of the neural tube, notochord and somites are found, also a postanal extension of the cloaca, the tail gut. At 12 mm., 6 weeks, the bud reaches its maximum length, 1 mm., with about 10 segments. Thereafter it regresses leaving only 3–5 coccygeal vertebrae in the adult.

Cloacal Membrane. Figs. 8 & 9

The cloacal membrane is a narrow, elongated structure of some thickness stretching from the tail bud to the connecting stalk. There is, as yet, no anterior abdominal wall between the stalk and the membrane.

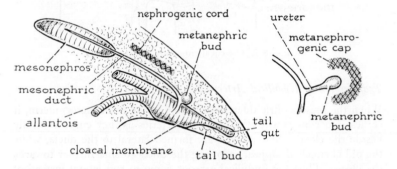

8.9. Components of the metanephros.

Metanephric Bud. Fig. 9

As the mesonephric duct turns ventrally to enter the cloaca it gives off a diverticulum, the metanephric (ureteric) bud, the first indication of the adult kidney. The stalk of the bud soon narrows to give the ureter. The intermediate cell masses belonging to the lower lumbar and sacral regions join together to form a continuation of the nephrogenic cord, the cells of which condense to cover the bud as the metanephrogenic cap. At a later stage the bud, cap and surrounding mesenchyme will build the kidney.

Limb Buds. Fig. 10

The limbs arise as thickenings of the body wall, the fore- and hindlimb buds, consisting at first only of undifferentiated mesenchyme with a covering layer of ectoderm. The ectoderm covering the early limb bud is especially thickened (see also figs. 9.10 and 9.14). This ectoderm is

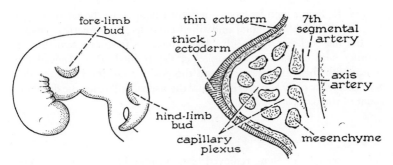

Fig. 8.10. Fore- and hind-limb buds.

carried away from the trunk as the limb grows and eventually forms the epidermis of the hairless skin of the palm, sole, nail field and dorsum of the terminal phalanx, while the thinner ectoderm drawn in at the base of the bud forms hairy skin. The mesenchyme is vascularized by a capillary plexus supplied by an axis artery arising, in the case of the fore-limbs, from the seventh cervical intersegmental artery and for the hind-limb from the fifth lumbar artery. The fore-limb is, throughout its development, a little in advance of the hind-limb.

Determination. Fig. 11

The mesenchyme of the early limb bud is, on ordinary microscopical examination, quite undifferentiated. Yet if a bud is isolated and grown in a hanging drop of nutrient fluid a relatively normal limb develops. In spite of the absence of a circulation bone formation is more or less normal. Even fragments of this mesenchyme cut out of the embryo and grown separately differentiate into recognizable structures, the upper

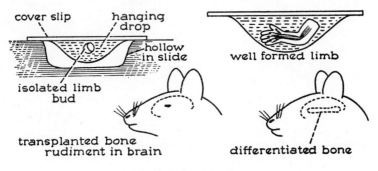

Fig. 8.11. Isolated fragments.

end of the femur with its trochanters for example, or the region of the knee joint with the femoral and tibial condyles, but not their shafts. So the mesenchyme is already 'determined' to form one structure or another, and consists of a 'mosaic' of determined areas. Indeed the bones with their main features and the positions of the joints are determined very early and precisely, and themselves influence the development of the parts around them.

Speed of Growth. Fig. 11

A bone only 1 mm. long removed from a rat embryo can be grown in the brain of an adult rat to 10 mm. It reaches this length in the same time as does a similar bone growing in its normal situation. So it appears that not only the shape of an organ, but also its rate of growth may be already determined in the early embryo.

9

Sections of 6 and 7 mm. Pig Embryos (30 days)

The Serial Sections

Pig embryos are easily obtained. Their foetal membranes are very different from those of man, but the embryos themselves are, at 7 mm. or about one month old, similar. A 7 mm. embryo cut at 10 μ gives about 700 sections. If these are mounted each on a separate slide a large class can all study one embryo. Some sections from the actual embryo being studied may be photographed and discussed before the class as a guide, but for a real understanding each student must study slides for himself. He should not be satisfied until he has drawn, and labelled in pencil as well as he can, at least one section from each of the following levels: cranial ganglia, branchial arches, heart, liver, mesonephros and somites, corresponding roughly to figs. 3, 4, 8, 10, 13 and 16, and at least one longitudinal section from an embryo comparable to fig. 17. If he has time for more, or for the study of older embryos (younger embryos are difficult to obtain and, in the pig, twisted and difficult to understand) so much the better, but a careful study of a few sections is more valuable than a glance at several. After (and not before) labelling, each drawing should be discussed with a demonstrator.

Section Through the Crown of the Head. Figs. 1 & 2

As shown in the index diagram a section through the crown cuts the hindbrain, for the midbrain and forebrain are bent out of the plane of section at the cephalic flexure. Over most of its extent the wall of the neural tube consists of three layers, an ependymal layer next to the cavity of the tube, a mantle layer with crowded cells hardly distinguishable from the ependymal layer, and a marginal layer consisting of cell processes only without nuclei, and so appearing clear in the section. The brain is surrounded by a vascular plexus.

The head is covered by an ectoderm one or two cells thick, separated from the neural tube by loose mesenchyme. The embryo is twisted at this stage, so that the section is not perfectly transverse and more mesenchyme appears on one side than on the other. In the mesenchyme lie small blood vessels, and, on one side only, the otocyst and accessory nerve.

The embryo lies in the amniotic cavity which is separated by the amnion from the general space filling most of the chorionic vesicle, the extraembryonic coelom. The amnion is a thin two-layered membrane, with an inner layer of amniotic ectoderm and an outer of primary mesoderm, each a simple layer of flattened cells.

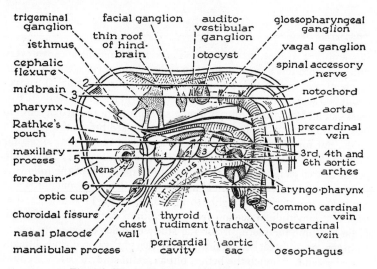

Fig. 9.1. The levels of figs. 2-6 are indicated.

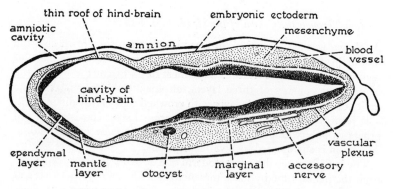

Fig. 9.2.

Section Through the Otocysts and Nuclei of the Cranial Nerves.
Figs. 1 & 3

The neural tube has two expanded parts, the diamond-shaped hind-brain and the circular midbrain, separated by the narrow isthmus. The

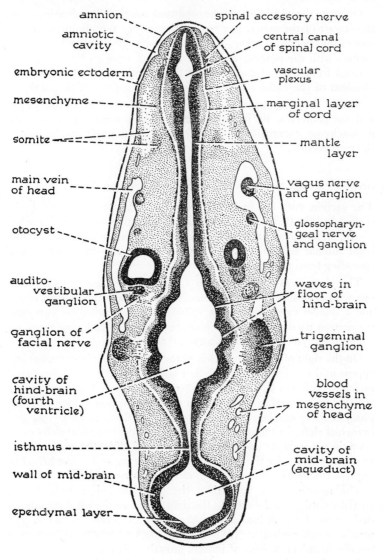

amnion

spinal accessory nerve

amniotic cavity

central canal of spinal cord

embryonic ectoderm

vascular plexus

mesenchyme

marginal layer of cord

somite

mantle layer

main vein of head

vagus nerve and ganglion

otocyst

glossopharyn- geal nerve and ganglion

audito- vestibular ganglion

waves in floor of hind-brain

ganglion of facial nerve

trigeminal ganglion

cavity of hind-brain (fourth ventricle)

blood vessels in mesenchyme of head

isthmus

cavity of mid-brain (aqueduct)

wall of mid-brain

ependymal layer

Fig. 9.3.

hindbrain is continued into the spinal cord, and the spinal accessory nerve lies on either side of the cord. The hindbrain is cut near its floor where its walls are thrown into conspicuous folds. The otocysts appear as a pair of thick-walled vesicles lying behind the widest part of the hindbrain.

Opposite the widest part of the hindbrain lies the trigeminal ganglion, the largest of the series, attached to the wall of the hindbrain by the root of the trigeminal nerve. The audito-vestibular ganglion lies in close contact with the otocyst, and immediately in front of this, since it is derived from the same segment of the neural crest, lies the sensory ganglion of the facial nerve, the geniculate ganglion of the adult.

Behind the otocyst two more ganglia, those of the glossopharyngeal and vagus nerves, lie in the mesenchyme. A large vein drains the mesenchyme, eventuallly joining the precardinal, which will be seen in the next section.

Section Through the Pharynx and Branchial Arches. Figs. 1 & 4

The section cuts the forebrain with Rathke's pouch applied to its surface and the two internal carotid arteries passing on either side of the pouch. At its other end the section cuts the spinal cord with its alar and basal laminae separated by the limiting sulcus, and the anterior horn spreading from the basal lamina.

The pharynx is cut twice, with a larger cavity where it is continuous with the mouth, for the stomodeal membrane has broken down, and a smaller cavity nearer the spinal cord. Between these two parts the section passes through the pharyngeal floor, cutting the aortic sac where the thyroid rudiment is attached to its wall.

On the right of the embryo (left of the drawing) the branchial arches appear in the order labelled 1, 2, 3, 4 and 6. The first three pharyngeal pouches, A, B and C, are seen between the arches, A and C continuous with the pharyngeal cavity and B cut off in this section but joined to the pharynx more cranially. A is clearly open to the exterior by the first branchial cleft, for though in human embryos the branchial clefts never open, in pig embryos they often do. B is closed off by the 2nd branchial membrane and C by the 3rd. The pouches correspond to the branchial grooves labelled on the other side of the drawing.

The mouth has a projection, the part labelled 'mouth cavity', partially cutting the 1st arch on the right, while on the left it completely divides the 1st arch into a maxillary and a mandibular process. The processes and arches are made of condensed mesenchyme such as is usually found in regions of intense growth activity.

From the aortic sac springs the 3rd aortic arch of the left side, while on the right it appears separate, lying in the interior of the branchial arch. The 4th aortic arch is separate on the left, but on the right is joining the right aorta. A precardinal vein lies lateral to the aorta on either side.

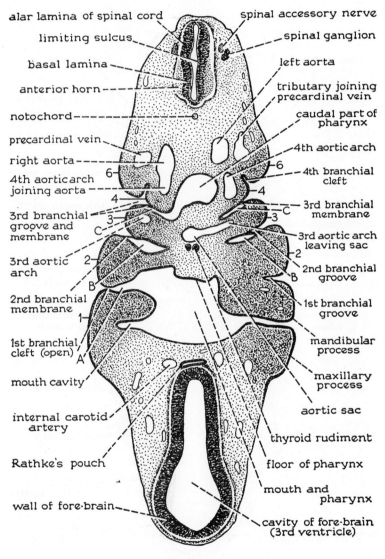

alar lamina of spinal cord

spinal accessory nerve

limiting sulcus

spinal ganglion

basal lamina

left aorta

anterior horn

tributary joining precardinal vein

notochord

caudal part of pharynx

precardinal vein

4th aortic arch

right aorta

4th branchial cleft

4th aortic arch joining aorta

3rd branchial membrane

3rd branchial groove and membrane

3rd aortic arch leaving sac

3rd aortic arch

2nd branchial groove

2nd branchial membrane

1st branchial groove

1st branchial cleft (open)

mandibular process

mouth cavity

maxillary process

internal carotid artery

aortic sac

Rathke's pouch

thyroid rudiment

floor of pharynx

wall of fore-brain

mouth and pharynx

cavity of fore-brain (3rd ventricle)

Fig. 9.4.

Section Through the Eyes and 6th Arches. Figs. 1 & 5

Each eye springs from the forebrain by an optic stalk which expands into an optic cup. The stalk and cup are hollow, enclosing a cavity which is a continuation of the cavity of the forebrain or 3rd ventricle.

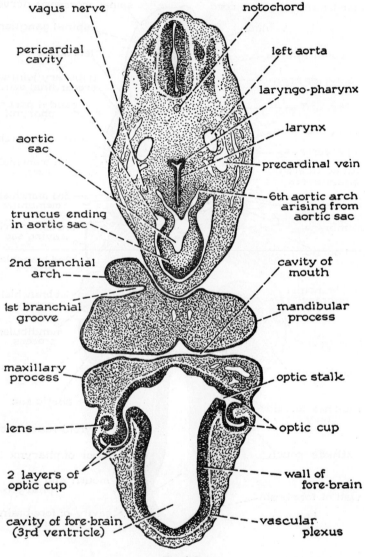

vagus nerve

notochord

pericardial cavity

left aorta

laryngo-pharynx

aortic sac

larynx

precardinal vein

6th aortic arch arising from aortic sac

truncus ending in aortic sac

2nd branchial arch

cavity of mouth

1st branchial groove

mandibular process

maxillary process

optic stalk

lens

optic cup

2 layers of optic cup

wall of fore-brain

cavity of fore-brain (3rd ventricle)

vascular plexus

Fig. 9.5.

On the right the section includes the lens vesicle, still not completely separated from the overlying ectoderm. On this side the optic cup appears to have no floor as the section has passed through the choroidal fissure.

The mouth is wide and separates the ends of the mandibular processes from the maxillary processes. The 2nd branchial arch is cut tangentially but the remainder do not appear in the section (see index diagram).

The section cuts the cranial end of the pericardial cavity containing the thick-walled truncus as it ends in the aortic sac. From the sac spring the right and left 6th aortic arches. These diverge round the larynx which has a thick epithelium and a narrow lumen, compressed from side to side.

Section Through the Auricles and Nasal Placodes. Figs. 6 & 7

The spinal cord and spinal nerves are particularly well shown in this section. The cord is divided into dorsal and ventral parts by the limiting sulcus. In the dorsal part the mantle layer is thickened on either side to form the dorsal horns, joined by a thin roof plate. The ventral horns are developed as very thick bulging masses with densely packed cells near the central canal, and more loosely packed cells peripherally, the two horns joined by a thin floor plate. The dorsal and ventral roots of the spinal nerves spring from the cord and join. In the dorsal roots lie the dense spinal ganglia.

The oesophagus and trachea lie in the median plane, one dorsal to the other, surrounded by a common condensation of the mesenchyme that will later form their muscular and cartilaginous coats. Lateral to these are the right and left aortae and the cardinal veins. It would be difficult to decide from this section alone whether they were pre-cardinal or postcardinal, but examination of more cranial and caudal sections of the series shows that in fact the veins are joining here to form the common cardinal veins. This is confirmed by the presence of a distinct pleuro-pericardial fringe projecting into the coelom where the left vein lies nearest. This fringe will help later to close off the pericardium.

The heart shows the two thin-walled auricles separated by the very thin septum primum. The left ventricle is cut tangentially through its spongy wall, and the thick-walled truncus is seen springing from the right ventricle. Between these structures lies the loose tissue of the septum intermedium, but this is better shown in the next section. The pericardial cavity can be traced right round the heart in this section, for there is a passage, the transverse sinus, between the trachea and the auricles. The cavity is enclosed by the thin chest wall.

The section cuts the free end of the forebrain and the mesenchyme surrounding it. The ectoderm, thin elsewhere, is greatly thickened on either side to form the nasal placodes, now dipping into the mesenchyme to form the nasal pits.

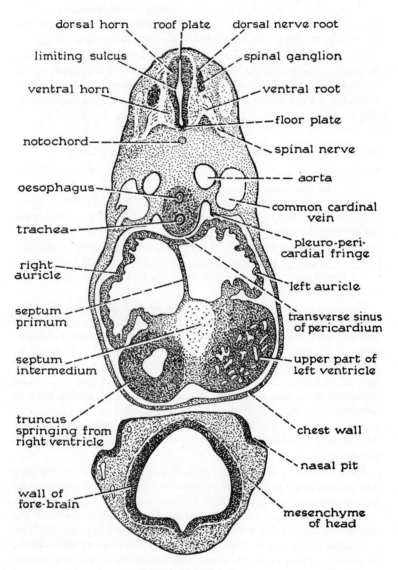

Fig. 9.6.

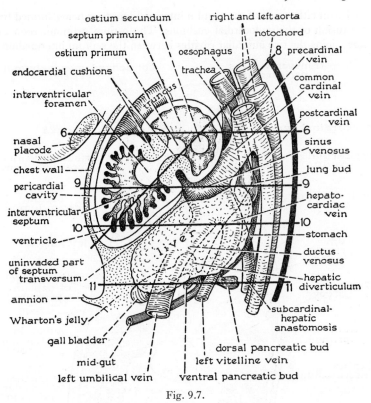

Fig. 9.7.

Section Through the Four Main Chambers of the Heart. Figs. 7 & 8

In this section, taken from a series cut in a different plane, as shown in the index diagram, the heart presents a pattern of diagrammatic simplicity with the midline septa and openings shown particularly well. It comes from the first serial sections of an embryo prepared in Iraq (by Mr. Saeed Kellow).

The two auricles are separated by the septum primum. This has a main part which is thin, pierced by the ostium secundum, and a thick soft edge approaching, but not yet fused with, the septum intermedium. The auricles are separated from the ventricles by the deep coronary groove, but there are no coronary vessels at this stage, for the heart wall is still thin enough to be supplied from the blood within.

The ventricles have a thick spongy wall, as in the adult frog, and their cavities are separated by the thick interventricular septum. Right and left auriculo-ventricular passages lead from the auricles to the ventricles on either side of the septum intermedium, now being formed by the fusion of the endocardial cushions. The ostium primum, soon to close, joins the two auricular cavities between the septum intermedium

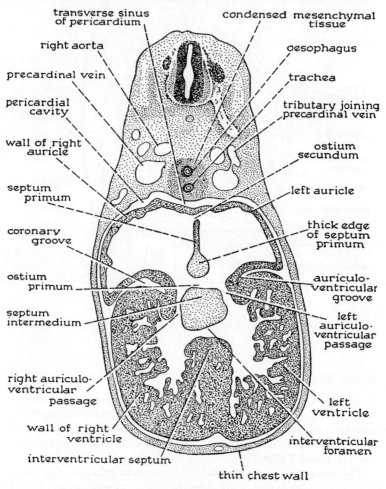

transverse sinus of pericardium

condensed mesenchymal tissue

right aorta

oesophagus

precardinal vein

trachea

pericardial cavity

tributary joining precardinal vein

wall of right auricle

ostium secundum

septum primum

left auricle

coronary groove

thick edge of septum primum

ostium primum

auriculo-ventricular groove

septum intermedium

left auriculo-ventricular passage

right auriculo-ventricular passage

left ventricle

wall of right ventricle

interventricular foramen

interventricular septum

thin chest wall

Fig. 9.8.

and the free edge of the septum primum. The interventricular foramen joins the two ventricular cavities between the septum intermedium and the interventricular septum. This never closes, for it is used later to join the ascending aorta to the left ventricle.

Though the ventral part of this section passes more caudally than in the section described previously, its dorsal part passes more cranially.

So it cuts the precardinal vein before it reaches the common cardinal. The spinal cord and a spinal nerve are again well shown, but left un-labelled. Name the parts for yourself.

Section Through the Lung Buds and Sinus Venosus. Figs. 7 & 9

The right and left aorta are approaching each other preparatory to joining and near each vessel a rudiment of the autonomic system, later forming a sympathetic ganglion, can be made out. Ventro-lateral to the aorta is a cardinal vein. From this relation alone it would be difficult

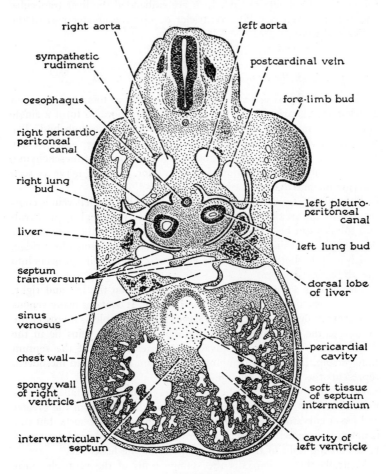

Fig. 9,9,

to distinguish it as pre- or postcardinal, but since this section is cut at the level of the lung buds it must be postcardinal (see index diagram, fig. 7).

The oesophagus is unchanged but the trachea has ended in the two lung buds. Each bud consists of a thick-walled diverticulum surrounded by a layer of mesenchyme. It projects into the medial wall of a C-shaped cavity, the pericardio-peritoneal canal (pleuro-peritoneal canal of some writers), a coelomic space which will later form the pleural cavity. The dorsal lobe of the liver lies beside the lung bud.

The septum transversum is seen as a band of tissue stretching across the embryo from side to side. In it are embedded the liver trabeculae and the sinus venosus. The ventricles occupy the greater part of the pericardial cavity.

Section Through the Stomach and Liver. Figs. 7 & 10

So far the aorta has appeared on each side, but at this level the right and left vessels are uniting ventral to the notochord to form a single channel. From this springs the axis artery to the fore-limb bud. The bud consists of an ectodermal covering several layers deep and specially thickened at the margin of the bud, and a dense mass of mesenchyme, without visible differentiation, in its interior. A large nerve is seen entering the bud to form the brachial plexus.

The right postcardinal vein is well developed but the left vein is compressed by the cranial end of the mesonephros. Mesonephric glomeruli and tubules can be recognised, but with difficulty as they are already degenerating at this end of the mesonephros. The mesonephric ridge, which includes both mesonephros and post-cardinal vein, projects into the abdominal coelom.

The liver is a large organ equally developed on the left and right, which crosses the whole width of the abdomen. It contains many venous spaces of various sizes. One of the larger is recognisable as it crosses the midline as the ductus venosus, and the hepato-cardiac veins lie in the dorsal lobes. The liver has invaded the greater part of the septum transversum, but a layer of uninvaded septum transversum mesoderm represents the future diaphragm.

Dorsal to the liver lies the stomach, sectioned as it falls onto its right side. The originally dorsal and ventral borders form the greater and lesser curvatures. A mesogastrium attaches it to the aorta, but this is not yet expanded to form a greater omentum. A C-shaped peritoneal recess on the right of the stomach represents the lesser sac.

Ventrally the section cuts the spongy walls of the ventricles near their tips, and passes through the base of the interventricular septum.

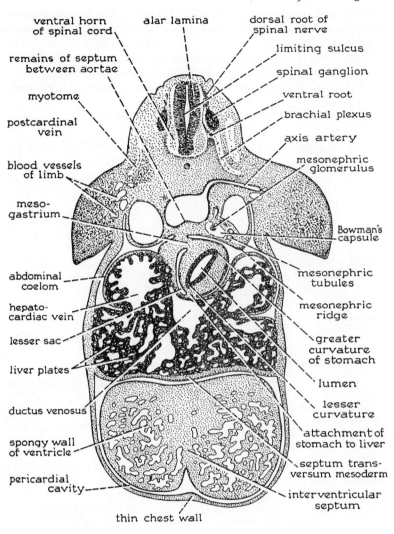

ventral horn of spinal cord

alar lamina

dorsal root of spinal nerve

remains of septum between aortae

limiting sulcus

spinal ganglion

myotome

ventral root

postcardinal vein

brachial plexus

axis artery

blood vessels of limb

mesonephric glomerulus

meso-gastrium

Bowman's capsule

abdominal coelom

mesonephric tubules

hepato-cardiac vein

mesonephric ridge

lesser sac

greater curvature of stomach

liver plates

lumen

ductus venosus

lesser curvature

spongy wall of ventricle

attachment of stomach to liver

pericardial cavity

septum trans-versum mesoderm

interventricular septum

thin chest wall

Fig. 9.10.

Section Through the Pancreatic Buds and Inferior Vena Cava.
Figs. 7 & 11

Here, at its maximum development, the mesonephros is a large organ projecting far into the abdominal coelom. Along its medial side is a group of glomeruli, each surrounded by its capsule. The main bulk is made up of tubules. At the ventrolateral corner lies the mesonephric duct, similar in structure to a tubule but distinguished by its position and its relation to a venous space lying nearby. Dorsolaterally the postcardinal vein is in intimate relation with the tubules. The mesonephros is supplied directly from the aorta by mesonephric arteries.

The liver is fading out, and is seen as three separate parts, each

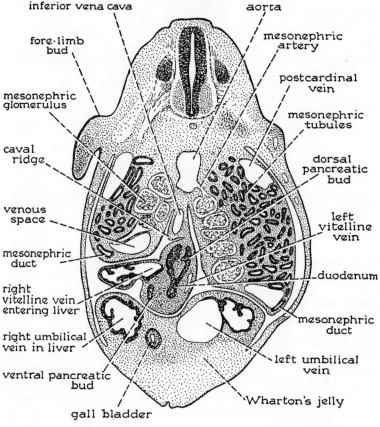

Fig. 9.11.

associated with an entering vein. The right and left umbilical veins from the chorion have approached the liver, running in the Wharton's jelly of the abdominal wall, and are now ending in the liver substance, as they do in the pig, though not in the human embryo.

The large lumen of the stomach has narrowed to the small lumen of the duodenum. A two-horned mass of cells, the dorsal pancreatic bud, projects dorsally from the duodenum into the dense mesenchymal tissue surrounding it, and a smaller mass, the ventral pancreatic bud, lies to the right of the duodenum. The gall bladder lies embedded in the anterior abdominal wall between the umbilical veins as they enter the liver.

The segment of the inferior vena cava formed by the subcardinal-vitelline anastomosis projects into the peritoneal cavity as the caval ridge.

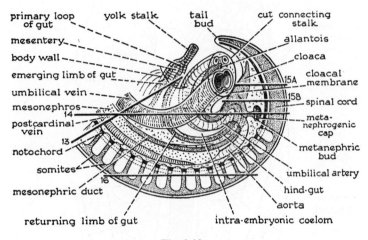

primary loop of gut   yolk stalk   tail bud   cut   connecting stalk

mesentery — allantois

body wall — cloaca

emerging limb of gut — 15A cloacal membrane

umbilical vein — 15B

mesonephros — spinal cord

14 — meta-nephrogenic cap

postcardinal vein — 13

notochord — metanephric bud

somites — umbilical artery

16 — hind-gut

mesonephric duct — aorta

returning limb of gut   intra-embryonic coelom

Fig. 9.12.

Section Through the Communication Between Intra- and Extra-embryonic Coeloms. Figs. 12 & 13

In this section, cut on a different plane to most of the others (see index diagram), the mesonephros appears as before. But now the subcardinal veins are seen side by side near the midline, ventral to the aorta. The aorta has a rectangular appearance, for it is giving off intersegmental arteries dorsally and mesonephric arteries ventrally. A condensed mass of cells near the aorta represents a rudiment of the autonomic system.

Between the right and the left mesonephros is attached the mesentery, a relatively thick sheet of connective tissue which projects ventrally into the extra-embryonic coelom. The gut is cut three times, as shown in the index diagram. At the root of the mesentery it appears as the duodenum with the left vitelline near it, and nearer the free margin as the primary loop.

The abdominal wall is thick, swollen by a mass of Wharton's jelly in which the two umbilical veins are passing towards the liver. Between

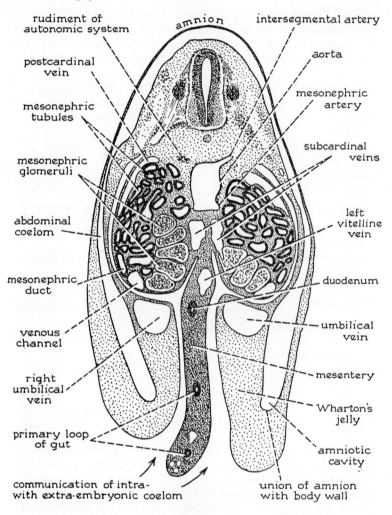

rudiment of
autonomic system

amnion

intersegmental artery

postcardinal
vein

aorta

mesonephric
tubules

mesonephric
artery

mesonephric
glomeruli

subcardinal
veins

abdominal
coelom

left
vitelline
vein

mesonephric
duct

duodenum

venous
channel

umbilical
vein

right
umbilical
vein

mesentery

primary loop
of gut

Wharton's
jelly

amniotic
cavity

communication of intra-
with extra-embryonic coelom

union of amnion
with body wall

Fig. 9.13.

the mesentery and the wall the intra- and extra-embryonic coeloms
communicate freely. The umbilical veins pass cranially lateral to the
coelom, the vitelline veins medial to it. In the early embryo they meet
in the septum transversum, but now that the liver has invaded the
septum the two streams meet in the liver.

The amnion is attached to the abdominal wall on each side, closing
off the amniotic cavity.

Section Through the Mesentery and Metanephric Buds. Figs. 12 & 14

This section catches the curl of the embryo so that the spinal cord and notochord are cut twice and appear at both ends of the drawing. The

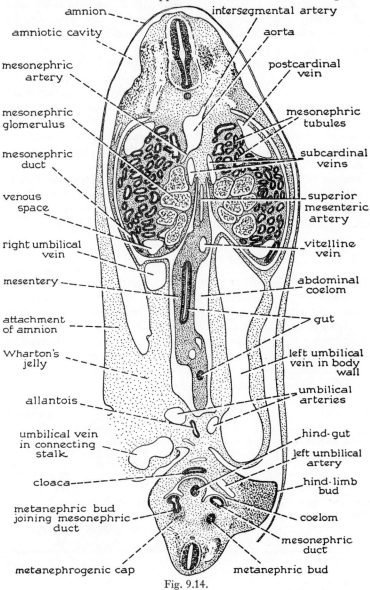

amnion

amniotic cavity

mesonephric artery

mesonephric glomerulus

mesonephric duct

venous space

right umbilical vein

mesentery

attachment of amnion

Wharton's jelly

allantois

umbilical vein in connecting stalk

cloaca

metanephric bud joining mesonephric duct

metanephrogenic cap

intersegmental artery

aorta

postcardinal vein

mesonephric tubules

subcardinal veins

superior mesenteric artery

vitelline vein

abdominal coelom

gut

left umbilical vein in body wall

umbilical arteries

hind-gut

left umbilical artery

hind-limb bud

coelom

mesonephric duct

metanephric bud

Fig. 9.14.

postcardinal vein, much compressed by the mesonephros more cranially, is here again expanding into a large vessel. The two subcardinal veins are again well seen.

The mesentery is in full development with the gut cut twice, once longitudinally and once transversely (see fig. 12). The superior mesenteric artery appears in its root and will run along the axis of the primary loop.

The hind-limb bud resembles the fore-limb but is smaller, and the nerves and vessels are not yet recognisable. The hind-gut and the mesonephric ducts are approaching the cloaca. On the right side of the embryo the metanephric bud, with its metanephrogenic cap, is seen joining the duct; on the left it is separated in the section. Between the hind-limb bud and the coelom the left umbilical artery is passing to the connecting stalk. The allantois with the umbilical arteries lying close on either side and the umbilical veins nearby, together with the surrounding Wharton's jelly, constitute the connecting stalk, cut here at its attachment to the embryo. The left umbilical vein is seen coursing along the abdominal wall, while the right vein is cut twice, once as it enters the wall and again as it nears the liver.

Two Sections Through the Cloaca. Figs. 12 & 15

The spinal cord is seen as a simple tube, for the limiting sulcus and horns have not yet developed in this region. The posterior continuations of the aorta have fused into a middle sacral artery and the postcardinal veins can still be recognised. The cloaca has a thick endodermal wall and a narrow lumen and is separated from the exterior, represented by the amniotic cavity, by the cloacal membrane. A mesonephric duct enters the cloaca on either side. The dorsal part of the cloaca which will form the rectum has not yet been cut off.

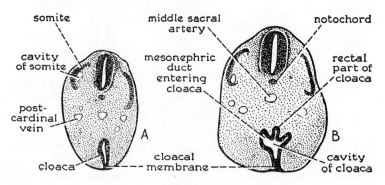

Fig. 9.15.

Section Through the Hinder End of the Embryo. Fig. 16

The section cuts the spinal cord longitudinally. A series of somites lies on either side, with the intersegmental vessels traversing the mesenchyme between them and the cord. A few of the somites have a temporary cavity, but this disappears. The embryo is surrounded by the two-layered amnion.

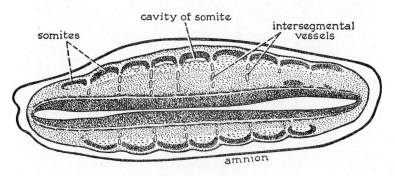

Fig. 9.16.

Longitudinal Section. Fig. 17

As the embryo is twisted it is impossible to cut a section which follows the median plane of the embryo throughout its length. Thus the section chosen cuts the central nervous system three times, once through the brain and twice through the spinal cord. The notochord is also cut three times and the aorta twice.

The brain is well shown, bent sharply at the cephalic flexure. The forebrain lies close to the heart, with the hemisphere forming an outgrowth from its dorsal surface, and Rathke's pouch applied to its ventral surface. It passes into the midbrain without any sharp boundary, but the isthmus separates the midbrain from the hindbrain with its thin roof.

The stomodeum is partly separated from the pharynx by the remains of the stomodeal membrane. The aortic sac lies in the floor of the pharynx with the thyroid rudiment close to its wall. Caudally the pharynx passes into the laryngopharynx and this is seen dividing into the trachea and oesophagus. A lung bud lies at the level of the liver.

The sinus venosus, one auricle, one ventricle and the truncus are well shown. The sinus is recognised by its thin wall and position just cranial to the liver, the auricle by its thin wall, relatively large size and relation to the pericardial cavity, the ventricle by its thick spongy wall

E

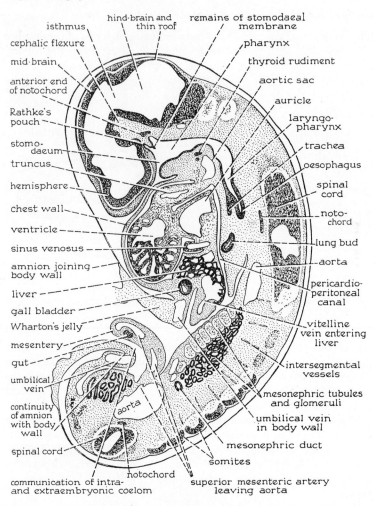

Fig. 9.17.

and the truncus by its thick but smooth wall and its position. The liver has developed in the septum transversum and a vitelline vein is seen entering it. The gall bladder lies on its caudal surface. In this region, at the attachment of the amnion and connecting stalk, the mesenchyme of the body wall becomes thick and loose textured, forming Wharton's jelly. The umbilical veins are cut as they make their way in the body wall towards the liver.

One mesonephros has been cut longitudinally showing a series of glomeruli and tubules, the other obliquely, and the mesentery springs from between them. The mesentery carries the gut and contains the superior mesenteric artery. The abdominal coelom is continuous ventrally with the extraembryonic coelom, and cranially it can be followed dorsal to the liver into one of the pericardio-peritoneal canals and so to the pericardial cavity. The lung bud indents the canal.

A series of somites is seen in the more caudal part of the section, with intersegmental vessels traversing the mesenchyme between them. The amnion encloses the amniotic cavity in which the embryo lies.

SECOND PART

7mm. STAGE TO BIRTH

Face and Nose

Branchial Arches and Cranial Nerves. Fig. 1

At 7 mm. the nasal placode, the eye and otocyst are all developing, the branchial arches are formed, and the 1st arch is already divided into a maxillary and a mandibular process. The trigeminal ganglion gives off its three main divisions, ophthalmic, maxillary and mandibular, the first passing cranial to the eye and the others into the two processes. The facial, glossopharyngeal and vagal ganglia supply the 2nd, 3rd and remaining arches respectively.

Early Face. Fig. 1

The eyes are far apart, on the sides of the head, and the nasal placodes, now sinking into the underlying mesenchyme towards the forebrain to form the nasal pits, are also well separated. The mesenchyme covering the forebrain with its diverticula, the eyes and cerebral hemispheres, is

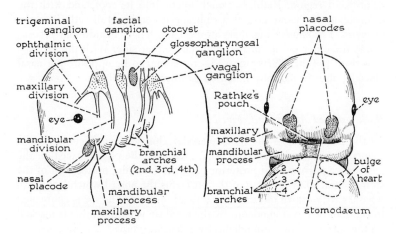

Fig. 10.1. The branchial arches.

now growing rapidly to provide a material for the formation of the face
The region between the eyes and between the nasal placodes is known
as the frontonasal process though it is never as clearly defined as are
the maxillary and mandibular processes.

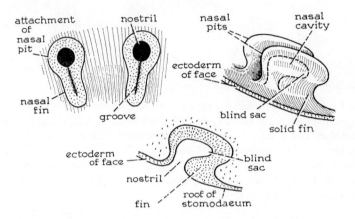

Fig. 10.2. The racket-shaped attachment of nasal pit.

Stomodeum. Fig. 1

The stomodeum is a wide shallow depression poorly defined from the
frontonasal region. Rathke's pouch opens into its roof, and with the
destruction of the stomodeal membrane it continues directly into the
pharynx. The shorter maxillary process and longer mandibular process
define the outer corner of the mouth on either side. With the forward
growth of the maxillary process its tip comes to lie opposite the nasal
placode, already partially indented to form the nasal pit.

Racket-shaped Attachment. Fig. 2

As the head expands the nasal placodes elongate and become racket-
shaped. An expanded anterior part, including the pit, is placed on the
face, and a narrow solid posterior part, the nasal fin, on the roof of the
stomodeum. The opening of the pit forms the nostril which leads into
a nasal cavity drawn out caudally into a blind sac. The sac is connected
to the roof of the stomodeum by the nasal fin whose position is marked
on the surface of the stomodeum by a shallow groove. At this stage the
nasal cavity has no opening except the nostril.

Circumnasal Rim. Fig. 3

A circumnasal rim forms round each nostril, broken only at the attachment of the nasal fin. Laterally the rim ends in the alar process, medially in the globular process. The two rims are separated by the nasal field, a wide, depressed area. The mandibular processes soon reach the midline and fuse, defining the position of the lower lip, but the nasal field passes gradually into the roof of the stomodeum, so that the position of the upper lip is still undefined.

The Nose. Fig. 3

The eyes move from the side of the head onto the face, to allow binocular vision. The nostrils and circumnasal rims approach each other, so that the part of the nasal field between them is obliterated. The globular

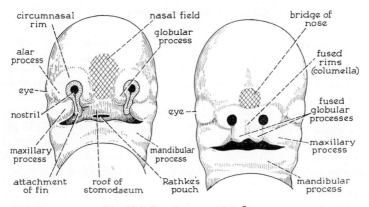

Fig. 10.3. Development of the face.

processes fuse to form the philtrum of the upper lip, and the rims the columella. The alar processes persist as the alae of the adult nose, and the remains of the nasal field form the bridge.

Fusion Across the Fin. Figs. 3 & 4

The part of the fin near the nostril thins and disappears, probably by death and disintegration of its cells. The mesenchymes of the maxillary and globular processes, no longer separated by the fin, fuse. For a time a shallow groove marks the former position of the fin, but soon this too is lost. The nostril is now completely separated from the caudal part of the fin by tissue which will later contribute to the upper lip and jaw.

The caudal part of the fin is still attached to the roof of the stomodeum, with a shallow groove marking the line of attachment.

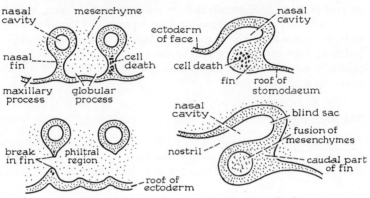

Fig. 10.4. Fusion of the mesenchymes across the fin.

Hare Lip. Fig. 5

Should the part of the nasal fin near the nostril persist, the mesenchymes of the maxillary and globular processes will fail to fuse at the end of week 6. The fin ectoderm between them eventually breaks down, leaving a cleft between the nostril and mouth, so giving a hare lip, which may be bilateral. Once the proper time (7th week) is past there is no hope of fusion of the mesenchymes, and the union must be completed surgically. Failure on both sides gives bilateral hare lip with an isolated median element derived from the globular processes.

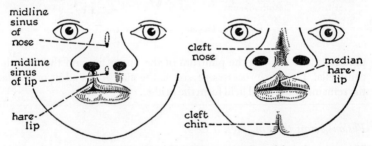

Fig. 10.5. Hare lip and midline anomalies of face

Congenital Midline Anomalies. Fig. 5

Normally as the nasal rims and globular processes fuse the ectoderm of the nasal field between them is buried and destroyed. If they fail to fuse a midline furrow, giving cleft nose or median hare lip, both rare anomalies, results. If they fuse but without complete destruction of the epithelium, buried skin may result. This may become glandular and secretory, forming closed cysts or sinuses opening on the face. Partial failure of fusion of the mandibular processes is relatively common, resulting in cleft chin, which most students will find among their acquaintances.

Nerves of Face. Fig. 6

The maxillary and mandibular divisions of the trigeminal nerve enter the corresponding processes early in development (fig. 6.7), and are carried with the processes as they spread. When the mesenchymes of the maxillary and globular processes fuse, branches of the maxillary division spread from their own process into the philtrum. In bilateral hare lip they cannot do so, and branches of the ophthalmic division take their place, while the facial vessels reach the element along the bridge of the nose. It is clear that though the nerves and vessels may give indications of the mode of development of a region, there is no precise correspondence between embryonic and neurovascular territories.

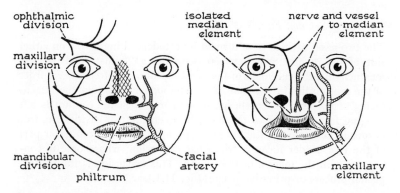

Fig. 10.6. Sensory nerves and vessels in normal face and in bilateral hare lip.

The Posterior Naris. Fig. 7

The caudal part of the nasal cavity, the blind sac, expands rapidly at the expense of the nasal fin. The fin becomes short and its remains are stretched to form the bucco-nasal membrane, which now separates the nasal cavity from the roof of the stomodeum. Eventually the membrane breaks down, giving a posterior naris. At this stage the nasal

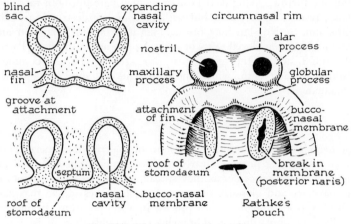

Fig. 10.7. The bucco-nasal membrane

cavity resembles that of the adult frog, having two openings, a nostril opening on the face and a posterior naris on the roof of the stomodeum.

The Nasal Septum. Figs. 8 & 11.1

The two nasal cavities were formed by ectodermal ingrowths on either side of the midline, but between them lies an undisturbed mass of mesenchyme, the rudiment of the nasal septum. The free lower border of the septum lies in the roof of the stomodeum between the attachments of the fins. When the posterior nares open the septum is left hanging from the roof of the nose, between the two cavities, its free border in contact with the developing tongue. The most anterior part of the free border forms a triangular expansion, whose base is continuous with the margin of the mouth. This expansion is the median palatal process or primitive palate. (fig. 11.1.).

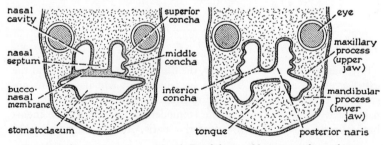

Fig. 10.8. Coronal sections of head. Breakdown of bucco-nasal membrane.

The Nasal Cavity. Fig. 8

Even before the break in the bucco-nasal membrane projections of the lateral wall of the nasal cavity indicate the superior and middle conchae. The inferior concha is formed at the attachment of the membrane. It would at this stage be possible to look from the stomodeum directly into the nasal cavity on either side of the septum.

The Three Parts of the Nasal Pit

It should now be clear that the originally continuous racket-shaped attachment of the nasal pit behaves differently in its three parts. The expanded facial part remains open as the nostril. The intermediate part, narrowed to form the fin, is destroyed by death of its cells and fusion of the adjacent mesenchymes. The caudal part is at first narrow, but is later expanded to form the bucco-nasal membrane, which eventually breaks down to form the posterior naris.

The Mouth

The Palatal Processes. Figs. 1 & 2

The nasal cavity is cut off from the mouth by the development of three palatal processes, one median and two lateral. The median is small and grows caudally from the fused globular processes. The lateral processes are large and grow medially from the maxillary processes towards the nasal septum. They cannot reach the septum at this stage as the tongue is in the way, so their margins become turned down at the sides of the tongue.

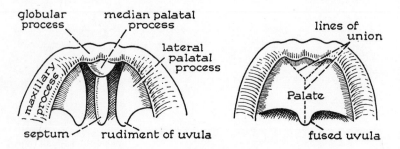

Fig. 11.1. The palatal processes.

The Palate. Fig. 2

The tongue now withdraws from the septum and the palatal processes swing up into a horizontal position so that their margins come into contact with each other and with the margin of the nasal septum. A tripartite fusion between these three structures completes the hard palate and cuts off the nasal cavities from the mouth. A continuation of the fusion between the palatal processes beyond the septum forms the soft palate. The uvula appears as a pair of small projections placed one on each lateral palatal process before fusion.

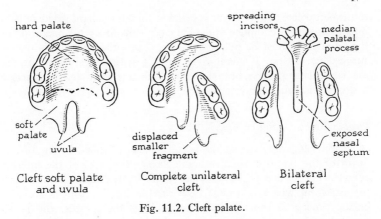

Fig. 11.2. Cleft palate.

Cleft Palate. Fig. 2

One or both palatal processes may fail to reach the midline so that an opening remains between the nasal cavity and the mouth. This opening interferes with sucking, allows food to enter the nasal cavity and seriously impairs speech. The spreading of the processes towards the midline is due to the rapid production of muco-polysaccharide in the ground substance by the fibroblasts of its mesenchyme. This occurs from week 7 to week 10, a relatively long period and later than the critical times for most other organ formations. So cleft palate is usually found without anomalies of other regions, sometimes the result of cortisone administered to the mother that has crossed the placental barrier. The cleft may affect the uvula only, cleft uvula, or be confined

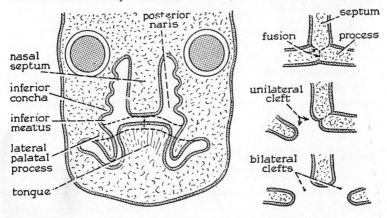

Fig. 11.3. Palatal processes, coronal sections.

to the soft palate and uvula, or continue through the hard palate to pass into a hare lip. A complete unilateral cleft leads to a gross distortion of the jaw which must be put right by an appropriate prosthesis before surgical correction. Bilateral cleft gives a symmetrical jaw, but the median palatal process becomes displaced anteriorly and the incisor teeth, if they erupt, project forward.

Oral and Nasal Cavities. Fig. 3

It will be remembered that the posterior naris, formed by the break-down of the bucco-nasal membrane, lay at the level of the inferior concha. The palatal fold is attached more ventrally, to the maxillary process from which it grows. Between the inferior concha and the palatal process is the inferior meatus. Thus the meatus, which was originally part of the stomodeum is added to the nasal cavity.

Choanae and Nasopharynx. Fig. 4

The tripartite fusion of the palatal processes and nasal septum involves only a part of the septum. Caudally, as the soft palate, the processes turn away from the septum to divide the nasopharynx from the oro-pharynx. This leaves the caudal part of the septal margin free between the choanal openings from the paired nasal cavities to the nasopharynx. In the embryo this free part of the margin is placed horizontally, but later, in the adult, becomes vertical. At the time of birth it still is more horizontal than vertical.

Labio-gingival Lamina. Fig. 5

So far the mouth has been bounded by a series of processes, maxillary, globular and mandibular. These form a thick rounded margin represent-ing the as yet unseparated lip (labium) and gum (gingivus). Two sheets of epithelium, the dental lamina and the labio-gingival lamina, grow into the mesenchyme, their position indicated on the surface by a shallow labio-gingival groove. Later the groove deepens, separating the lip from the gum.

Dental Lamina. Figs. 4 & 5

The dental lamina, concerned with tooth formation, grows into the tissues of the gum. From it are formed a series of enamel organs, one for each deciduous tooth, and later one for each permanent tooth. The mesenchyme of the gum condenses beneath each enamel organ to form a dental papilla. Once the series of organs is complete the lamina itself disappears.

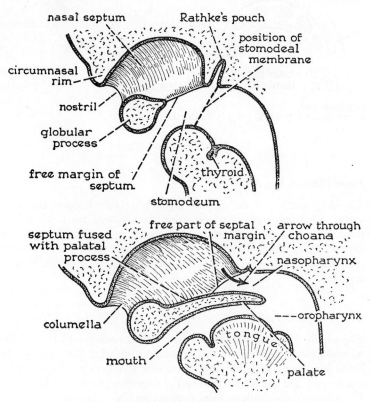

Fig. 11.4. The nasal septum and the choana.

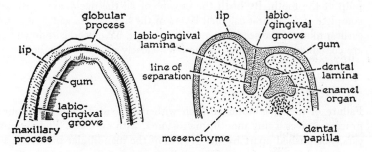

Fig. 11.5. The labio-gingival groove and lamina.

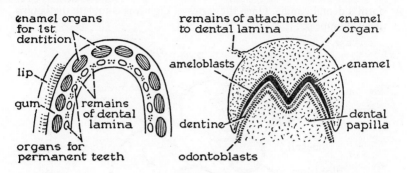

enamel organs
for 1st
dentition

lip

gum

remains
of dental
lamina

organs for
permanent teeth

remains of attachment
to dental lamina

ameloblasts

dentine

odontoblasts

enamel
organ

enamel

dental
papilla

Fig. 11.6. The enamel organs.

Enamel Organ. Fig. 5

The deep surface of the enamel organ takes the shape of the tooth surface with indentations representing the cusps of the tooth it is to form. The cells of this surface, ectodermal in origin since they are derived from the epithelium of the original stomodeum, become arranged as a columnar epithelium, the layer of ameloblasts (enamel formers).

Dental Papilla. Fig. 5

The papilla fits into the enamel organ and its surface cells become the odontoblasts, which control the formation of dentine. Now the ameloblasts lay down layer after layer of enamel, the earliest layer becoming the deepest and the latest forming the superficial surface of the tooth. The odontoblasts similarly form dentine, but here the outermost layer is the earliest and the later layers are laid down within it, gradually reducing the size of the papilla, whose remains eventually form the pulp of the tooth. By birth the crowns of all the deciduous teeth are complete and the first calcified layers of the first permanent molar, the earliest tooth of the second dentition to erupt, have been laid down. All are contained in the bones of the jaws, closely crowded together.

Microstoma. Fig. 7

Failure of the first branchial arch with its maxillary and mandibular processes gives a tiny mouth, microstoma, without teeth. The external ears, not pushed apart by the growth of the mandibular processes, lie close together in the front of the neck.

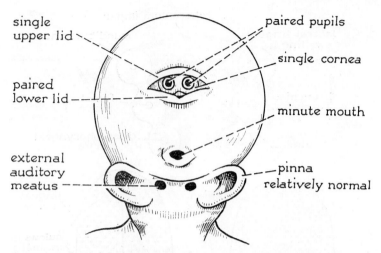

single
upper lid

paired pupils

single cornea

paired
lower lid

minute mouth

external
auditory
meatus

pinna
relatively normal

Fig. 11.7. Combined anomalies of the head, cyclopia, microstoma and misplaced external ears. Based on Keith's classic specimen, *Brit. med. J.* 1909, reproduced in Willis, Borderland of Embryology and Pathology, 2nd Ed. 1962.

The Tongue. Fig. 8

A lateral lingual swelling grows from each 1st arch and a median swelling, the tuberculum impar, appears between them. These three eminences fuse to give the anterior two-thirds of the tongue. The posterior third is built up from the medial ends of the 2nd arches and the pharyngeal floor between them. The sulcus terminalis follows the boundary between the 1st and 2nd arch derivatives and the foramen caecum marks the position of the thyroid rudiment. In the early embryo the sulcus is bowed forwards, but as the anterior part of the tongue expands it comes to make a V with the apex pointing towards the pharynx.

Nerves of the Tongue. Fig. 8

The anterior two-thirds of the tongue receive a common sensory supply from the lingual nerve, a branch of the mandibular, the nerve of the 1st arch (the chorda tympani subserving taste). But the posterior third is upplied by the glossopharyngeal with no contribution from the facial, though 2nd arch tissue is included in this third. Here, as in the face, the arrangement of sensory nerves is a poor guide to embryonic development. The tongue muscles are mentioned under head myotomes in Chapter 20.

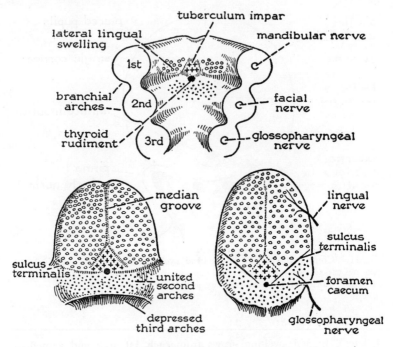

Fig. 11.8. The tongue.

Salivary Glands. *Fig. 9*

The development of glands in general may be illustrated from that of the parotid gland, the first of the salivary glands to develop. The epithelium of the mouth forms a solid outgrowth into the mesenchyme of the cheek and later extends towards the developing ear. Here it

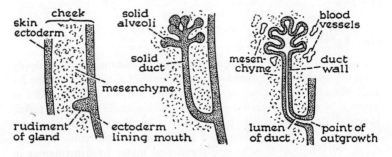

Fig. 11.9. A salivary gland.

breaks into branches, the terminal twigs swelling to form acini. The whole system now hollows out, giving a lumen in each branch and acinus. The mesenchyme between the acini forms the stroma of the gland and the surrounding mesenchyme condenses to form the capsule. Thus eventually the mouth epithelium comes to be represented only by the epithelium lining the acini and ducts of the glands while the mesenchyme forms the remainder of its structure. The original outgrowth forms the main duct (compare the kidney and its ureter), and the point of outgrowth is marked by the point of entry of the duct into the mouth. The embryological development of most exocrine glands can be deduced accurately from a consideration of their adult structure.

Branchial Arches: Their Vessels, Skeleton, Muscles and Nerves

Arterial Arches. Fig. 1

At 7 mm. the 1st and 2nd aortic arches have already degenerated while the 3rd, 4th and 6th arise from the aortic sac and traverse the corresponding branchial arches. A cranial prolongation of the dorsal aorta forms the internal carotid. From the 6th arch a new vessel, the pulmonary artery, develops to supply the rapidly enlarging lung bud. From these beginnings the carotid (3rd arch), aortic (4th arch) and pulmonary (6th arch) systems will develop.

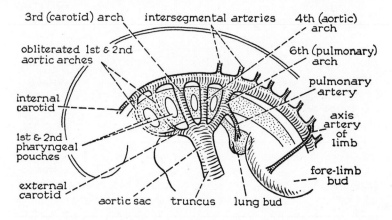

Fig. 12.1. The aortic arches.

Axis Artery. Fig. 1

The limb buds arise after the intersegmental arteries have been formed, and the original axis artery of the fore-limb is derived as a branch from the seventh intersegmental. But with the increase of bulk of the bud the axis artery comes to include the stem of the intersegmental, so that the axis artery now springs directly from the aorta (fig. 9.10.). The proximal part of the axis artery becomes the subclavian of the adult.

Carotid System. Fig. 2

The section of aorta connecting the 3rd and 4th arches disappears, freeing the carotid system from the aortic. The original 3rd arch now forms the proximal part of the internal carotid, while the distal part is formed from the aorta. The external carotid is formed from the roots of the 1st and 2nd arches, and the common carotid from the aortic sac.

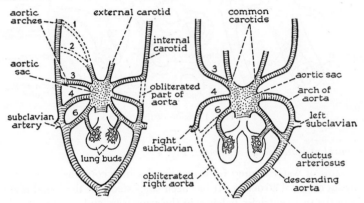

Fig. 12.2. Aorta and pulmonary artery.

Aortic and Pulmonary Systems. Fig. 2

On the right side the dorsal aorta is cut off from the pulmonary system by the closure of the distal part of the 6th arch. The 4th arch, a part of the aorta and the original subclavian artery now all form parts of the right subclavian, which at this stage arises directly from the aortic sac. The remainder of the right aorta disappears. On the left the 4th arch is preserved in the arch of the adult aorta, and the aorta of the embryo is included in its descending part. The distal part of the left 6th arch is preserved till birth as an open channel, the ductus arteriosus, connecting the pulmonary and aortic systems. On this side there are no additions to the embryonic subclavian artery.

Division of the Truncus. Fig. 3

The truncus is divided by a pair of spiral ridges. These ridges join giving two passages, the pulmonary trunk and ascending aorta, winding round each other. The ridges extend into the aortic sac, so that the pulmonary trunk is connected to the 6th arches and the aorta to the other vessels arising from the sac. Originally the ridges involve only the endothelium and the cardiac jelly, but eventually the muscular coat of the truncus also divides so that the aorta and pulmonary trunk are completely separated.

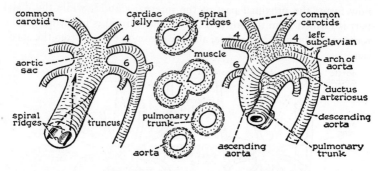

Fig. 12.3. Division of the truncus.

Control of Vascular Development. Fig. 4

Streams of water issuing from two nozzles may combine into a single stream of spiral structure. It has been suggested that the blood issuing from the two ventricles behaves in the same way, and it is the shape of the combined stream that decides the shape of the spiral ridges in the truncus. Again if, in the area vasculosa of a chick embryo, two vessels of equal size are chosen, in one of which blood is flowing fast and in the other slowly, then in a few hours time the vessel with the more rapid flow is found to have widened as compared with the other. Indeed the

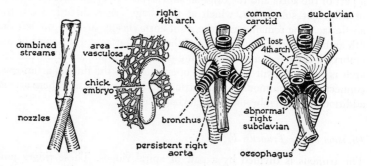

Fig. 12.4. Vascular development and anomalies.

development of the whole vascular system seems to depend on local conditions of blood flow through it and many anomalies of the heart probably arise through unequal division of the blood stream because endocardial cushions are badly formed.

Anomalies of the Aorta. Fig. 4

Occasionally the right aorta persists as well as the left, as is normal in the frog. There is then an arterial ring round the oesophagus and trachea which may tighten with development. This causes difficulty in respiration and swallowing and needs surgical correction. Loss of the 4th arch on the right side leads to an abnormal origin of the right subclavian from the descending aorta. The artery has to pass dorsal to the oesophagus and right bronchus and here again may cause pressure symptoms.

Branchial Skeleton. Fig. 5

For some time each branchial arch consists only of an epithelial covering, a mesenchymal core, a nerve and an arterial arch in the mesenchyme. But later a condensation in the mesenchyme forms the skeletal structures of the arch. Parts of this condensation go on to form cartilage which may later ossify, while other parts become fibrous or disappear.

Skeleton of the First Arch. Fig. 5

The most dorsal part of the skeleton of the 1st arch forms the incus, with the malleus separated by a joint. The malleus has a handle and an

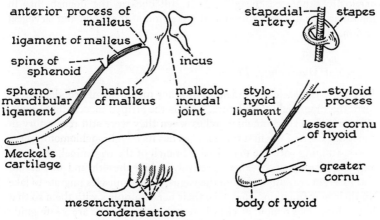

Fig. 12.5. Skeleton of the mandibular and hyoid (2nd) arches,

anterior process. The middle segment is represented by a ligament and the most distal by Meckel's cartilage which serves as a scaffolding round which a membrane bone, the dentary (mandible), is ossified from the mesenchyme. The middle, ligamentous, piece becomes attached to the sphenoid bone, and so subdivided into the anterior ligament of the malleus and the sphenomandibular ligament. A reduced skeleton of the lower jaw produces micrognathus or even agnathus.

Stapes. *Fig. 5*

The stapes is formed in the mesenchyme of the dorsal part of the 2nd arch. In the embryo it is pierced by the stapedial artery, and this artery is preserved throughout life in some animals, such as the mouse, but in most it disappears, leaving only the opening in the stapes. The auditory apparatus is so delicately adjusted that once it is developed it cannot grow, so all three ossicles of the middle ear, malleus, incus and stapes, are preformed in cartilage at their full size in embryonic life and do not enlarge after birth.

Styloid Process and Hyoid Bone. Fig. 5

The remainder of the 2nd arch skeleton is represented by the styloid process, stylohyoid ligament and the lesser cornu and upper part of the body of the hyoid. From the 3rd arch is formed the greater cornu and lower part of the body, completing the hyoid, The thyroid and cuneiform cartilages develop from the 4th and the cricoid, arytenoid and corniculate from the 6th arch. Thus the branchial skeleton, which originally supported the gills, as it still does in the dogfish, has in mammals become specialized as the skeleton of the hyoid and laryngeal regions.

Branchial Musculature. Fig. 6

Following the skeletal condensation other mesenchymal condensations form the branchial musculature. This is the striped muscle that originally controlled the branchial arches when they were still respiratory in function. Each of the first three arches has its own branchiomotor nerve fibres, running with the mandibular branch of the trigeminal, the facial and the glossopharyngeal, respectively, while the 4th and 6th arches and fore-gut are supplied from the vagus. As the muscles migrate to take up their new functions, they take their nerve supplies with them so that in the branchial region, as elsewhere, the motor nerves are a safe guide to the source of muscles.

Muscles of Mastication and Facial Expression. Fig. 6

The muscles of the 1st arch remain closely associated with the jaw, forming the four muscles of mastication, and also the mylohyoid, anterior belly of digastric and tensor tympani. Those of the 2nd arch mostly migrate onto the face, scalp (fronto-occipitalis) and neck, to form the muscles of facial expression and platysma, taking their particular

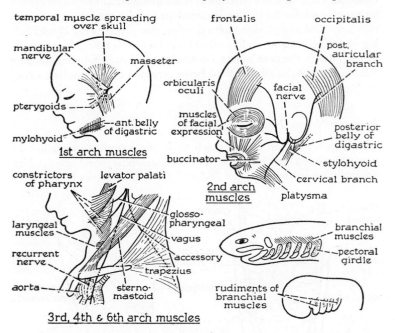

Fig. 12.6. Muscles of the branchial arches.

branches of the facial nerve with them as they go. A part of the 2nd arch musculature however remains in the hyoid region to form the posterior belly of the digastric, the stylohyoid and the stapedius.

Constrictors, Sternomastoid and Trapezius. Fig. 6

The remaining branchial musculature forms the palatal muscles, stylopharyngeus, constrictors of the pharynx, muscles of the larynx and striped muscle of the oesophagus, supplied by the glossopharyngeal and vagus nerves. But even in the dogfish a part of this musculature has transferred its attachment from the branchial cartilages to the shoulder girdle, and in man this part is represented by the sternomastoid and trapezius, with a special nerve, the spinal accessory, which has been split off from the original vagus to supply them.

Sucking and Swallowing

At 7 months (2 lbs. weight) sucking and swallowing movements are established and the foetus becomes viable, though the movements cannot be maintained for long. The suctorial pad of fat over the buccinator, relatively hard and of high melting point, that prevents the cheeks falling in during sucking, is, like other body fat at this time, almost totally lacking. But in the next month fat develops and swallowing of amniotic fluid becomes habitual, the fluid being returned to the amniotic cavity by the kidneys.

Recurrent Nerves. Fig. 7

The branchio-motor nerves reach their pharyngeal musculature by passing between the aortic arches. So the glossopharyngeal nerve, which originally passed between the 2nd and 3rd arches, comes to lie, in the

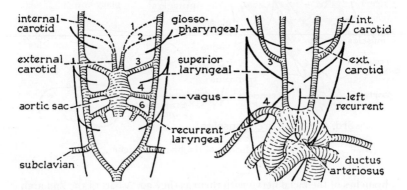

Fig. 12.7. Glossopharyngeal and vagus nerves.

adult, between the internal and external carotid arteries. The recurrent branch of the vagus, which originally reached the pharynx by hooking round the 6th arch, now hooks round the aorta beside the ductus arteriosus on the left and the subclavian on the right.

13

Pharyngeal Derivatives

Pinna and External Auditory Meatus. Fig. 1

At 5 mm. there is a complete set of branchial grooves separating the arches. Later the arches fuse so that the grooves become less prominent. On either side of the first groove, however, a series of small tubercles appears, three on the 1st, mandibular, arch and three on the 2nd, hyoid, arch. Later these fuse to form the pinna of the ear. A part of the epithelium of the first groove is hollowed out to form the external auditory meatus. At first the pinna lies near the midventral line. Failure to migrate dorsally may result in abnormal position, an extreme form being synotus, where the fused pinnae straddle the midline ventrally.

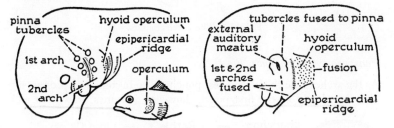

Fig. 13.1. The hyoid operculum.

Hyoid Operculum. Fig. 1

The 1st and 2nd arches are large and prominent, but the 3rd and 4th are much smaller and are sunk beneath the surface. Now a fold, the hyoid operculum, grows caudally from the 2nd arch, covering over the 3rd and 4th arches and burying them beneath the body surface. Finally the operculum fuses with the 6th arch, or epipericardial ridge, so as to leave a smooth surface which comes to form a part of the neck. In bony fishes the operculum is retained throughout life as a cover to the gills, and this was doubtless so in our fish ancestors.

Cervical Sinus. Fig. 2

In this way a considerable mass of ectoderm is buried in the neck, forming the embryonic cervical sinus. Usually this ectoderm is reduced to small closed cysts, and finally disappears completely. Sometimes it fails to do so, and a cyst or sinus may later form hair follicles and sebaceous glands, or may become infected, in either case necessitating surgical removal. A persistent cervical sinus may open on the surface of the neck, usually at the anterior border of the sternomastoid, or may open into the pharynx by perforation of a branchial membrane. Occasionally a continuous passage may be left between the pharynx and the exterior.

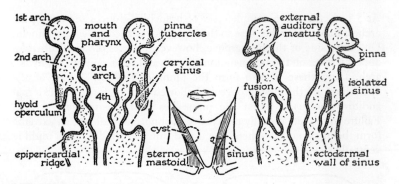

Fig. 13.2. The cervical sinus.

Pharyngo-tympanic Tube. Fig. 3

As the arches fuse the pharyngeal pouches also become less prominent. The dorsal part of the first pouch, however, becomes drawn out and

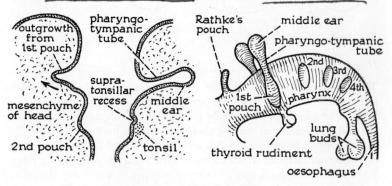

Fig. 13.3. The pharyngo-tympanic tube.

expanded at its free end to form the pharyngo-tympanic (Eustachian) tube and cavity of the middle ear.

Rudiments of Vestibulo-cochlear Apparatus. Fig. 4

The main structures of the vestibulo-cochlear apparatus are now laid down: The otocyst from general ectoderm, the ganglion of the 8th nerve from neural crest ectoderm, the pharyngo-tympanic tube and

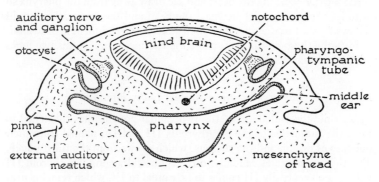

Fig. 13.4. Constituents of the auditory apparatus.

middle ear cavity from pharyngeal endoderm, the external acoustic meatus from remains of ectoderm of the first branchial groove and the pinna from tubercles of the 1st and 2nd branchial arches.

Tonsillar Fossa and Tonsil. Figs. 3 & 5

The second pharyngeal pouch is mostly obliterated. The underlying mesenchyme becomes filled with lymphoid tissue, forming the tonsil. The tonsil bulges into the fossa nearly obliterating it, leaving only a

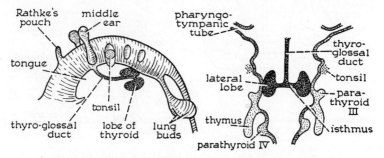

Fig. 13.5. Thyroid, thymus and parathyroid rudiments.

remnant, the supratonsillar recess of the adult. Ingrowths of the endoderm of the pouch make the tonsillar crypts.

Migration of the Thyroid. Fig. 5

At first the heart and pericardium lie immediately ventral to the floor of the pharynx as in the adult dogfish. The thyroid rudiment, growing from the pharyngeal endoderm, comes into close contact with the aortic sac (fig. 9.4.). As the heart and sac draw away from the pharynx so as to leave a mobile neck between the head and the thorax, the thyroid follows for part of the way. The solid stalk of the rudiment is drawn out to form the thyroglossal duct, while the rudiment itself forms two lobes joined by a narrow isthmus.

Parathyroid Glands and Ultimobranchial Body. Figs. 5 & 6

Meanwhile the endoderm of the two caudal pouches has thickened to form the rudiments of the parathyroids III and IV, named from their pouches of origin, and the thymus, which grows from the third pouch. The thymus migrates caudally to the thorax, dragging parathyroid III with it. So eventually III comes to lie caudal to IV, which remains near its place of origin. Thus the first parathyroid encountered by the thyroid on its migration is IV, and this eventually becomes embedded in the lateral lobe of the thyroid gland, while III lies more caudally in the mesenchyme of the neck. Other epithelial cells of the fourth pouch form, in some animals, a separate endocrine gland, the ultimobranchial, which secretes a calcitonin. In man these cells are for a time recognisable as the ultimobranchial body of the embryo, but later they become added to the parathyroid gland to form its 'C' cells, of similar function.

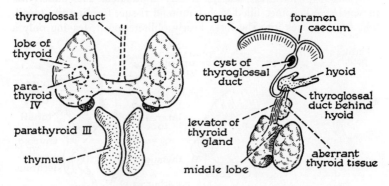

Fig. 13.6. Normal and abnormal pharyngeal pouch derivatives.

Thymus. Figs. 6 & 7

The two thymic masses come together in the thorax to form the adult thymus. But whereas in the parathyroids the original epithelial cells from the pharynx form the secretory cells which constitute the main bulk of the gland, in the thymus the epithelial cells are reduced to the corpuscles of Hassall. The main bulk is formed of lymphoid tissue derived from the mesenchyme. This tissue serves as a source of young lymphocytes (lymphoblasts) which pass, in the blood stream, to other lymphoid organs, such as the lymph nodes, spleen and Peyer's patches, where, under the influence of a hormone secreted by the thymus, they settle down and multiply. Removal of the thymus from newborn mice severely depresses lymphocyte production and the immunological reactions dependent on lymphocytes, while implantation of a thymus from another animal restores normality.

Thyroid Gland. Figs. 6 & 7

The solid mass of cells of the original thyroid rudiment becomes broken up by the ingrowth of mesenchyme carrying blood vessels. The mesenchyme forms the capsule and stroma of the gland and the epithelium the lining of the follicles. Colloid, secreted by the follicular epithelium, gradually accumulates in the cavity.

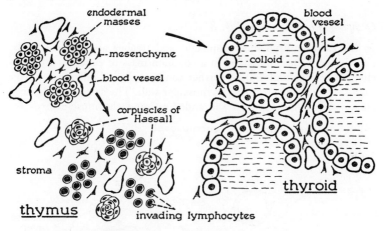

Fig. 13.7. Differentiation of thymus and thyroid.

Thyroglossal Duct. Fig. 6

The attachment of the thyroglossal duct to the floor of the pharynx persists as the foramen caecum, which becomes incorporated in the

F

tongue. The caudal end of the duct sometimes forms aberrant masses of thyroid tissue, or an extra, middle, lobe of the thyroid gland; or the mesenchymal coat of the duct may differentiate as a smooth muscle, the 'levator' of the gland. The rest of the duct usually disappears, its endodermal cells being destroyed. If the cells do persist they may become secretory and form a thyroglossal cyst, situated anywhere between the foramen caecum and the thyroid gland. The hyoid bone develops quite independently of the duct, which eventually becomes bent up behind the bone. Thus cysts may lie either in front of the bone or behind it, or by spread of the body tissue they may become enclosed within the bone.

History of the Thyroid

In our prochordate ancestors the thyroid gland produced a mucous secretion by which food particles entering the pharynx were trapped and swallowed. Thus the gland was an ordinary exocrine gland, opening into the pharynx by a duct. In the vertebrate embryo the duct is still represented by the solid thyroglossal duct, while the acini are modified to follicles. The anterior lobe of the pituitary, originally opening by a duct represented by the stalk of Rathke's pouch, had a similar history, becoming a ductless gland as the vertebrates developed jaws and came to chase larger prey.

Oesophagus and Trachea. Fig. 8

In the early embryo the oesophagus and trachea are represented by endodermal tubes embedded in a common mass of condensed mesenchyme (figs. 9.6 and 9.8). A further condensation of this mesenchyme round the oesophagus forms its muscular wall. The as yet undifferentiated mesenchyme between the endoderm and the muscle gives the

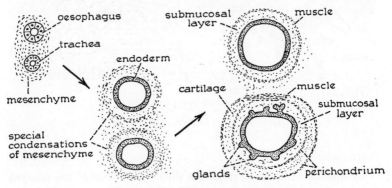

Fig. 13.8. Oesophagus and trachea.

submucosal layer. Round the trachea a similar condensation forms the cartilage and perichondrium, as well as the smooth muscle. The glands are developed later by epithelial outgrowths as in a salivary gland. The epithelial lining of both oesophagus and trachea is for a time ciliated, but in the oesophagus the cilia are later lost.

Tubular Structures

It is a general rule that in tubular structures the lining epithelium is developed first, and the layers of the wall are added later by differentiation of the mesenchyme. In the gut the endoderm forms the lining of the gut itself with the glands opening into it while the muscular and fibrous layers are added later. In the blood vessels the endothelial lining is formed first and later the musculo-elastic and connective tissue coats.

Laryngo-pharyngeal Orifice. Fig. 9

The orifice of the respiratory diverticulum is at first narrow and slit-like. Later three elevations appear at its margin, an epiglottis derived from 4th arch tissue and paired arytenoid swellings, making the orifice Y-shaped. The corniculate and cuneiform tubercles are derived from the arytenoid swellings.

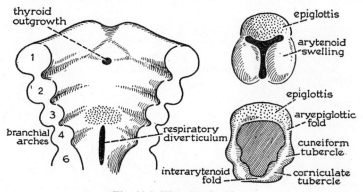

Fig. 13.9. The glottal orifice.

Tracheo-Oesophageal Communications. Fig. 10

A pinching-off process separates the trachea from the oesophagus. Sometimes a pair of lateral indentations separates the cranial parts of the tubes but they remain united below, giving a tracheo-oesophageal fistula. The fistula may persist till birth, but more usually the proximal part of the oesophagus loses its connection with the trachea and ends

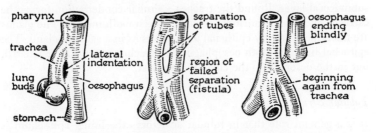

Fig. 13.10. Tracheo-oesophageal malformations.

blindly. In such cases the amount of amniotic fluid is excessive (hydramnios), probably because the foetus cannot swallow it, as it appears to do normally. Certainly, the swallowing reflexes are developed early, and can be elicited in the foetus by sweet tasting fluids.

The Lungs. Fig. 11

The endodermal lung bud subdivides to give a bronchial tree, whose ultimate subdivisions end blindly in the mesenchyme. The endoderm itself is a peculiar tall columnar epithelium with nuclei near the free surface. At this stage the lung has a glandular appearance.

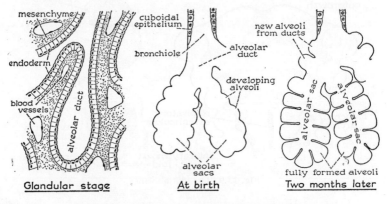

Fig. 13.11. Change of epithelium lining the lung.

The Alveoli. Fig. 11

The tree continues to branch, attaining about 27 dichotomies at the time of birth. In the final subdivisions the originally columnar and later cuboidal epithelium is stretched and thinned to give a cellular lining of adult type, allowing rapid molecular exchange through its substance. At birth the physiological units are the alveolar sacs, for the alveoli are hardly indicated, but in the first months of infancy the alveoli become fully developed as recesses from the sacs and new alveoli develop as expansions from the walls of the bronchioles.

Expansion of the Lungs

The mesenchyme is reduced to narrow septa which form the basement membrane for the epithelium and contain the collagenous, elastic and smooth muscular fibres, and the septal cells. The greater part of lung tissue is now made of capillaries. Before birth the alveolar sacs are relatively small and contain amniotic fluid swallowed by the foetus, but with the first breath they are expanded, and stay so throughout life. Possibly the expansion of the lung is helped by a vascular engorgement.

Surfactant

The first breath must be strong to overcome both the elastic resistance of the lung to expansion and the surface tension of the fluid coating the alveolar surfaces. A natural detergent, surfactant, is secreted by the alveolar wall and can be detected in lung extracts. It reduces the surface tension to a level where its expiratory effect is about equal to that of the elastic tissue. Some failures of the newly born to initiate or continue respiration are associated with lack of surfactant, especially likely in premature babies.

Viability of Foetus

With the flattening of the epithelium the lung becomes capable of efficient function and the foetus becomes viable. At this time, the beginning of the sixth month, the reflex mechanism of respiration is already developed. It should be possible to save some foetuses born even more prematurely by the transfusion procedures now being developed for animal foetuses.

14

The Later Heart

Transformation of the Veins. Fig. 1

The vascular plexus of the fore-limb empties by the subclavian vein into the precardinal, and this through the common cardinal into the sinus vensous. An oblique cross anastomosis is laid down joining the place of union of the subclavian and precardinal on the left to the right precardinal nearer the heart. The precardinals cranial to the union of

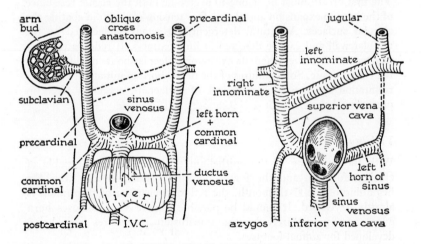

Fig. 14.1. The great veins.

the subclavians become the internal jugulars of the adult and the cross-anastomosis the left innominate. The superior vena cava is made up of sections derived from the embryonic precardinal and common cardinal, with the arch of the azygos vein, derived from the postcardinal, marking their extent.

Openings into the Sinus Venosus. Fig. 1

With the development of the cross-anastomosis, and the loss of the left postcardinal, little blood enters the left horn of the sinus venosus. So this horn becomes a relatively small channel. At this stage two large vessels, the superior and inferior venae cavae, and one small vessel, the reduced left horn, open into the main cavity of the sinus.

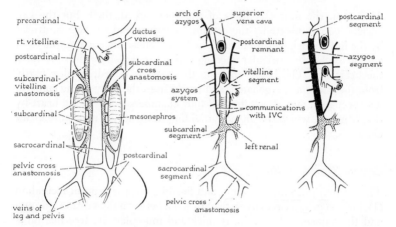

Fig 14.2. Completion of inferior vena cava and failure of subcardinal-vitelline anastomosis.

Completion of Inferior Vena Cava. Fig. 2

The veins of the hind limb bud at first drain into the posterior cardinal vein, at that time the only channel available. As the subcardinal veins become the main longitudinal channels of the abdomen the veins draining the legs and pelvic viscera drain into them by new sacrocardinal veins. A pelvic cross-anastomosis develops between the sacrocardinals so that all the blood can pass up the right vein. The inferior vena cava is now complete, consisting of four segments, from above downwards the vitelline stump, the subcardinal-vitelline anastomosis, the subcardinal segment and the sacrocardinal segment. If the pelvic cross-anastomosis fails to develop, both sacrocardinals persist giving a double inferior vena cava as far as the left renal vein.

Azygos System. Fig. 2

The stump of the right postcardinal vein forms the arch of the azygos vein but the main trunks of the azygos and hemiazygos are formed as longitudinal channels connecting the intercostal veins. The azygos and hemiazygos connect below with the inferior vena cava. Thus if the subcardinal-vitelline anastomosis fails to develop the blood from the legs and abdomen can reach the heart through a greatly enlarged azygos vein.

The Early Heart. See fig. 4.15

At 7 mm. the four chambers of the heart are distinct, but the two auricles communicate by the ostium secundum and the two ventricles by the interventricular foramen. The sinus venosus opens into the right auricle by the sinoauricular opening which is guarded by the only valve the heart has. The blood entering the heart is mixed, coming partly from the chorion, oxygenated, and partly from the body of the embryo, deoxygenated, and after filling all four chambers it leaves the heart by the truncus, which is already dividing into aortic and pulmonary channels.

Septum Secundum. Fig. 3

The septum primum (I) is thin and flexible, but the septum secundum (II), which grows caudally in the 8th week between the septum primum and the sinoauricular valve, is thick and muscular. Its free margin is curved and bounds an opening, the foramen ovale. When fully formed it covers the ostium secundum, but so long as the pressure in the right auricle is greater than that in the left the septum primum can be pushed aside so that blood can reach the left auricle. At birth the sudden influx of large quantities of blood from the lungs into the left auricle raises the pressure there and the flexible septum primum is pressed against the rigid septum secundum to close the foramen ovale.

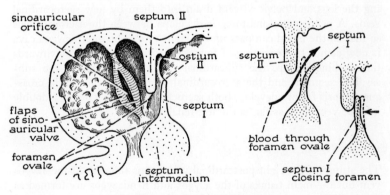

Fig. 14.3. The interauricular septa.

The Bulbus. Fig. 4

A new chamber, the bulbus, differentiates between the ventricles and the truncus. A pair of bulbar ridges, which continue the direction of the spiral ridges of the truncus, divide the chamber into two passages. The bulbar ridges finally fuse with each other and with the septum intermedium and interventricular septum. Thus the bulbus is divided so as to connect the left ventricle with the aorta and the right ventricle with the pulmonary trunk. The passage to the aorta makes use of the interventricular foramen, which thus remains open throughout life.

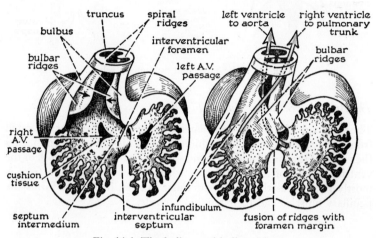

Fig. 14.4. The bulbus and bulbar ridges.

The Infundibulum. Fig. 4

As compared with the ventricle the bulbus is smooth within. The part connecting the right ventricle with the pulmonary trunk makes the infundibulum of the adult. A similar, though less developed, smooth region leading to the aorta makes the aortic vestibule of the left ventricle. The wall between the infundibulum and vestibule makes the non-muscular membranous part of the adult interventricular septum.

Aortic and Pulmonary Valves. Fig. 5

Between the spiral and the bulbar ridges, and in continuity with them, the cardiac jelly swells to raise a pair of valvular cushions, and another pair of cushions appears at the same level to complete a ring of four. As the aortic and pulmonary arterial channels separate so does the valvular region, the plane of division cutting one pair of cushions so that there are now six. Each cushion becomes condensed and re-shaped to a cusp, so that the aorta and pulmonary trunk are each provided with three valves.

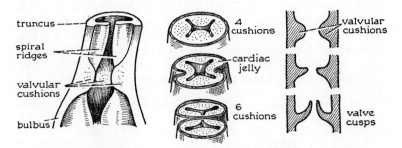

Fig. 14.5. The aortic and pulmonary valves.

The Auriculo-ventricular Valves. Fig. 6

The right and left auriculo-ventricular passages are, when first separated from each other, guarded by cushion tissue. This condenses and makes the cusps of the mitral and tricuspid valves. The fine spongework of the ventricular wall of the early heart, which resembles that of a frog, is mostly retained as the adult ventricular muscle, but a part forms the papillary muscles and their tendinous cords.

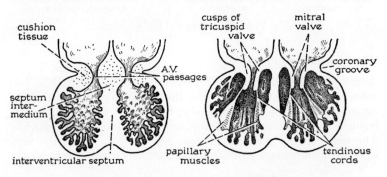

Fig. 14.6. The mitral and tricuspid valves.

Taking in of the Sinus Venosus. Fig. 7

As the new valves become functional the need for the old sinoauricular valve is lost. The opening between the sinus and right auricle widens so that the two chambers form one, the adult auricle, or, in other words, the sinus is taken into the auricle. The right common cardinal and the right vitelline vein, preserved as the superior and the inferior vena cava, now open directly into the right auricle, separated by an eminence, the intervenous tubercle. The original left horn of the sinus is drawn out to form the coronary sinus, while the left common cardinal vein dwindles to become the oblique vein of the left auricle. The left flap of the sino-auricular valve disappears, but the right is preserved as the so-called valves of the inferior vena cava and coronary sinus.

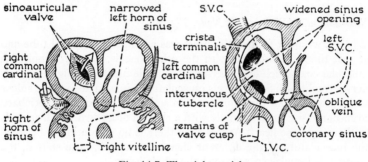

Fig. 14.7. The right auricle.

Left Superior Vena Cava. Fig. 7

Occasionally the oblique cross-anastomosis between the precardinal veins (see fig. 1) fails to develop. In this case, there are two superior venae cavae, a right and a left, as is normal in many mammals. The abnormal left vessel empties into the coronary sinus, since that sinus is the transformed left horn of the sinus venosus. The minute oblique vein of the left atrium normally marks the course the left cava would follow.

Taking in of the Pulmonary Veins. Fig. 8

The pulmonary veins at first enter the left auricle as a single vessel, but later the walls of his vessel are taken into the auricle, so that two, then

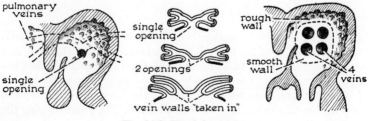

Fig. 14.8. The left auricle.

four, pulmonary veins enter separately. In both auricles the older embryonic part of the wall is rough, with pectinate muscles covering the inner surface, while the smooth part is made of tissues taken in later, the sinus venosus in the case of the right auricle and the pulmonary veins in the left. On the right the original groove between the sinus and auricle is still marked on the outside of the adult heart by the sulcus terminalis, and on the inside by the crista terminalis (fig. 6), in which the pectinate muscles end. On the left the precise boundary is lost.

Sino-auricular (Sinu-atrial) Node. Fig. 9

When it was first formed the heart was a simple endothelial tube and at that time it contracted with the other vessels. Later it acquired a muscular coat which took over the contractile function (long before any nerves reached it). Some of this musculature, situated deep to the sulcus terminalis between the sinus venosus and auricle, becomes specialized to form the sinoauricular node with a rhythm of its own which spreads to the rest of the heart. At this time there is no discontinuity between the musculature of the different chambers, so that, once started, the contractile impulse spreads over the whole heart.

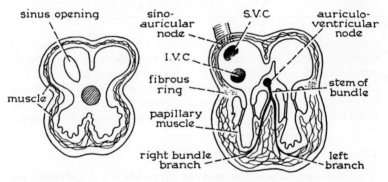

Fig. 14.9. The conducting system.

The Conducting System. Fig. 9

Much later the auricular muscle is cut off from the ventricular muscle by the fibrous ring of the heart. The impulse cannot pass this ring, so a conducting system has to be installed. A band of muscle persists between the auricle and ventricle to become the main stem of the bundle of His, and extensions on either side of the interventricular septum carry the impulse to the apex of each ventricle. The fibres forming the bundle, ordinary heart muscle at first, accumulate glycogen and enlarge to become Purkinje tissue. The bundle and its major branches are insulated from the general musculature of the heart by a connective tissue covering, but the terminal branches blend with the heart muscle. At the point of origin of the bundle a nodule of heart muscle differentiates to give the auriculo-ventricular node. Eventually the bundle forms the sole connection between the auricular and ventricular muscles.

The Foetal Circulation. Fig. 10

Oxygenated blood coming along the umbilical vein from the placenta reaches the heart mixed with blood from the foetus itself coming by the inferior vena cava. Entering the right auricle it is directed to the left by the intervenous tubercle, which lies between the superior and the inferior venae cavae, and by the persistent right cusp of the old sino-auricular valve. The thick edge of the septum secundum also helps to guide this stream of blood into the left auricle. Deoxygenated blood from the head and neck enters the right auricle by the superior vena cava. There is surprisingly little mixing of the two blood streams in the right auricle, so that when the auricles contract the oxygenated blood enters the left ventricle and the venous blood the right, and later these streams reach the aorta and pulmonary trunk. In foetal life, however, the lungs are contracted and little blood can circulate through them. But the ductus arteriosus provides a by-pass by which blood from the pulmonary trunk can reach the aorta.

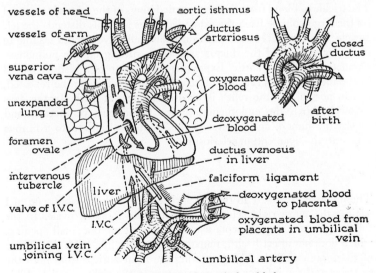

Fig. 14.10. The circulation before birth.

Distribution of the Streams. Fig. 10

The arch of the aorta is filled for the most part with arterial blood from the left ventricle, and this is distributed to the head and neck and upper limbs. The ductus arteriosus, bringing venous blood, enters the aorta just beyond its arch, so that its blood supplies the abdomen and lower limbs and, through the umbilical arteries, reaches the placenta. It is surprising that the lower half of the body has to make do, throughout its intra-uterine development, with poorly oxygenated blood, but such is the case. This may be a factor in the relatively slow development of this part of the body during prenatal growth.

Changes at Birth. Fig. 10

Sporadic breathing movements occur long before birth and amniotic fluid enters the lungs. At birth the respiratory centre, stimulated by impulses from the cold receptors of the skin and chemoreceptors sensitive to the state of the blood, induces the first breath. The alveolar sacs of the lungs open up and for the first time there is a free circulation of blood in their walls. They never completely empty again, so that it is possible to test whether a baby has lived by putting its lungs in water, to see whether they float or not.

Closure of the Ductus Arteriosus. Fig. 10

The trickle of blood from the unexpanded lungs of the foetus is replaced by the large volume from the expanded lungs and this distends the left auricle. Pressure receptors in the auricular wall serve a reflex leading to muscular closure of the ductus arteriosus, cutting off the pulmonary circulation from the aortic. Experimental inflation of a bladder placed in the foetal auricles sets off the reflex, and the effect can be repeated, but in normal life the reflex is used only once, at birth. In the days following birth the endothelium lining the ductus proliferates and blocks the lumen, and eventually the ductus is reduced to a fibrous remnant, the ligamentum arteriosum.

Closure of the Foramen Ovale. Fig. 3

The increased pressure in the left auricle presses the flexible septum I against the more rigid septum II, closing the foramen ovale. In the following weeks the two septa usually fuse, but often a small opening, which does not affect health, remains throughout life. With the closure of the foramen all the blood entering the right auricle from both the superior and the inferior vena cava passes to the right ventricle. This blood is now entirely venous, for the stream of arterial blood entering the foetus from the placenta is cut off when the umbilical cord is tied. The arterial and venous sides of the heart are now quite separate.

Coarctation of the Aorta. *Fig. 11*

The blood in the arch of the aorta goes to the head and arms, while the descending aorta gets its own supply through the ductus arterious. So, between the attachments of the subclavian artery and ductus, the flow is minimal in the foetus. After birth this section may narrow, causing coarctation. With the ductus closed, blood for the abdomen and legs must come through anastomoses between the subclavian arteries and those of the trunk. The most important are the anastomoses between the costal branch of the costocervical trunk supplying the 1st or 1st and 2nd intercostal spaces and the 3rd intercostal from the aorta,

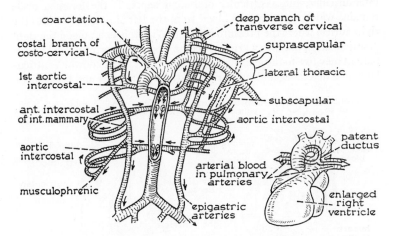

Fig. 14.11. Coarctation and patent ductus.

the anastomosis between the superior epigastric branch of the internal thoracic and the inferior epigastric of the external iliac, and the numerous smaller anastomoses of branches of the subclavian with the intercostal arteries. Through these channels sufficient blood reaches the abdomen and limbs to sustain life, but not enough for strong effort. The condition is diagnosable from the poor pulse in the foot as compared to the wrist, the X-ray shadow of the aorta and the notches made in the ribs by the enlarged and tortuous anastomoses. It can be treated surgically.

Patent Ductus Arteriosus. Fig. 11

If the ductus fails to close at birth it allows blood to pass between the systemic and pulmonary circulations. Since the pressure is greater in the aorta than in the pulmonary artery the shunt is from the arterial to the venous side. The large volume of blood passing through the relatively narrow ductus sets up a loud 'machinery murmur', diagnostic of the condition. The increased pressure in the pulmonary arteries leads to hypertrophy of the right ventricle, and may eventually cause a breakdown of compensation. The channel can be closed surgically.

Patent Foramen Ovale. Fig. 12

The heart has to develop, and at the same time to remain in full functional activity throughout prenatal life. Developmental defects are relatively common. Most are septal in origin, being due to failures of the interauricular, interventricular or bulbar septa. If the foramen ovale remains widely open arterial blood from the left auricle passes into the right auricle where the pressure is lower. With this extra blood the right ventricle and pulmonary artery become greatly hypertrophied but the patient may live to the age of 50 before the heart breaks down. Ventricular septal defects again lead to right-sided hypertrophy and breakdown.

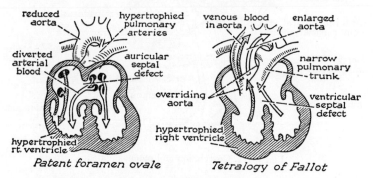

Fig. 14.12. Cardiac defects.

Tetralogy of Fallot. Fig. 12

The spiral ridges seldom fail to separate the aorta and pulmonary trunk, but may do so unequally so that the pulmonary trunk is poorly developed. The overdeveloped aorta cannot now be connected to the interventricular foramen, but comes to 'override' the interventricular septum and receives blood from both ventricles. As the two ventricles remain in open communication the pressure in the right ventricle is as high as that in the left and the wall of the right is hypertrophied. This combination of (1) ventricular septal defect, (2) narrowed pulmonary trunk, (3) overriding aorta and (4) right ventricular hypertrophy is the tetralogy of Fallot. The venous blood entering the aorta causes cyanosis, giving the greater number of 'blue babies'. Surgical treatment by formation of an artificial ductus arteriosus is often very successful. The pulmonary circulation is increased, pressure rises in the left ventricle, less venous blood enters the aorta and the baby becomes healthy and resumes normal growth.

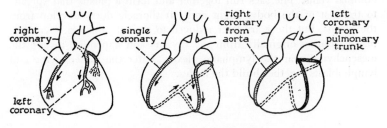

Fig. 14.13. The coronary arteries.

Coronary Arteries. Fig. 13

The early heart is sufficiently thin-walled not to need any special blood supply. But as the walls thicken a capillary plexus spreads over their surface from the aorta, and from this the coronary arteries are derived.

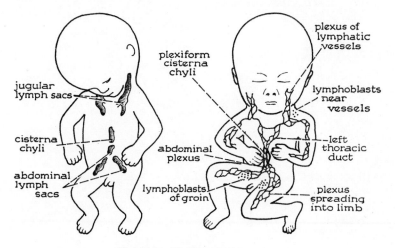

Fig. 14.14. The lymphatic system.

Sometimes only one is formed, an anomaly of no clinical significance. But an origin of the left artery from the pulmonary trunk leads to death in babyhood, for the venous blood cannot serve the heart.

Lymphatic Vessels and Nodes. Fig. 14

The lymphatic system appears as a pair of outgrowths of the internal jugular veins, the jugular lymph sacs. The cisterna chyli and abdominal lymphatics are developed from similar sacs sprouting from the abdominal veins. The sacs run together and form a plexus that spreads to the rest of the body. Right and left thoracic ducts develop from the plexus, but these are later reduced to a single channel. Lymphoblasts, coming originally from the thymus and appendix, invade the mesenchyme near the lymphatic plexus. Later the mesenchyme and lymphoblasts together build the nodes.

15

The Placenta and Twinning

The Decidua. Fig. 1

As the chorionic vesicle expands it encroaches on the uterine cavity and three parts of the endometrium become distinguishable. The part by which the embryo is attached to the wall of the uterus is the decidua basalis, the part invaginated into the uterine cavity is the decidua capsularis, and the part not directly related to the embryo is the decidua parietalis. The chorionic villi are embedded in the decidua basalis and capsularis. The embryo itself is closely surrounded by the amnion, and attached to its ventral surface are the yolk stalk and connecting stalk.

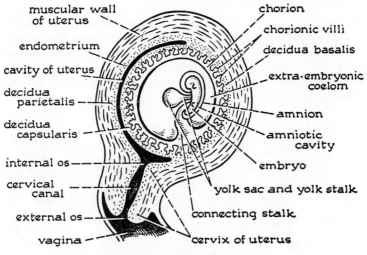

Fig. 15.1. Embryo in the uterus.

Rotation of the Embryo. Fig. 2

At first the dorsal surface of the embryo looks towards the decidua basalis. But later the embryo rotates, its head moving ventrally and its tail bud dorsally. It turns through two right angles, so that its dorsal surface faces away. At the same time the yolk sac and yolk stalk are brought near the connecting stalk.

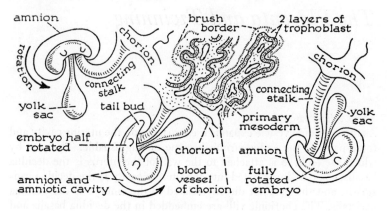

Fig. 15.2. Rotation of the embryo.

Expansion of the Amnion. Fig. 3

The amniotic fluid increases rapidly so that the amnion is expanded. By this stage the cloacal membrane is ruptured, and it is possible that most of the additional amniotic fluid is secreted as a urine by the mesonephros. As the amnion expands it presses the yolk stalk against the connecting stalk and eventually these three structures fuse to form the umbilical cord.

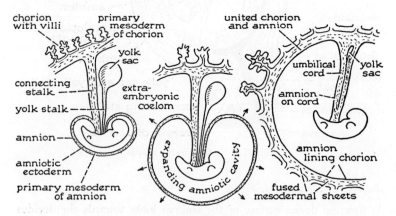

Fig. 15.3. Expansion of the amnion.

Obliteration of the Extra-embryonic Coelom. Fig. 3

The expansion of the amniotic cavity occurs at the expense of the extra-embryonic coelom. The coelom is reduced and finally obliterated, so that the amnion and chorion come into contact. The outer surface of the amnion and the inner surface of the chorion are both made of primary mesoderm, and these layers fuse, so that the amnion serves as a smooth lining to the chorionic vesicle. The foetus, floating in an abundant fluid at the end of a long flexible cord, can move more freely as its muscles develop. If there is too little amniotic fluid the foetus is likely to adhere to the inner surface of the amnion with resulting deformities.

Umbilical Cord. Fig. 4

The umbilical cord is developed from three originally separate components, the connecting stalk, the yolk stalk and the amnion. The connecting stalk traverses the extra-embryonic coelom and contains two umbilical arteries carrying blood from the foetus to the chorion, the umbilical vein bringing blood back to the embryo, and the allantois. The cord, when compacted by obliteration of the extra-embryonic coelom, thus contains three vessels and two endodermal tubes, the yolk stalk with its sac and the allantois. Its connective tissue, Wharton's jelly, has abundant ground substance and sparse fine fibres with a few cells, and is derived from primary mesoderm. The cord is surrounded by amniotic ectoderm.

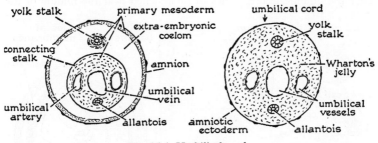

Fig. 15.4. Umbilical cord.

The Early Placenta. Fig. 5

In the early embryo the whole surface of the chorion is surrounded by well developed villi (fig. 1). But as the embryo grows the decidua capsularis becomes stretched and relatively thin, and its blood supply is impoverished. So the villi embedded in it regress, while the villi embedded in the thick, well vascularized decidua basalis grow. Eventually the villi become restricted to the region of the decidua basalis, and these form the bulk of the placenta, while the rest of the chorion becomes smooth. Later the embryo comes to occupy the uterus fully, and the decidua capsularis fuses with the decidua parietalis, obliterating the uterine cavity.

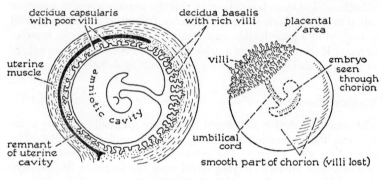

Fig. 15.5. Early placenta.

Attachment of Cord. Fig. 6

At the time of embedding the embryonic pole of the blastocyst, marked by the inner cell mass, usually leads the way. So the connecting stalk comes to be attached to the chorion opposite the decidua basalis, and later the umbilical cord comes to be attached near the middle of the

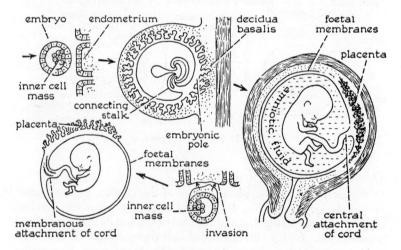

Fig. 15.6. Attachment of umbilical cord.

placenta. But sometimes a different part of the trophoblast comes into contact with, and invades, the endometrium. The location of the placenta is still decided by that of the decidua basalis, and the cord attachment by the inner cell mass. So the cord may be attached to the margin of the placenta (marginal attachment) or to the smooth part of the chorion (membranous attachment).

Spread of Cytotrophoblast. *Fig. 7*

Originally the syncytiotrophoblast with its lacunae lies in direct contact with the remains of the decidua. Later cell columns grow from the cytotrophoblast of the tips of the villi to reach the decidua, and spread between it and the syncytiotrophoblast, so the whole of the intervillous space comes to have a double lining, including both layers of trophoblast. These layers make the trophoblastic shell, which encloses the villi and intervillous space.

Placental Barrier. *Fig. 7*

Thus the foetal blood with its nucleated corpuscles is separated from the maternal blood by the placental barrier, which consists of layers of endothelium, primary mesoderm, cytotrophoblast and syncytiotrophoblast. Through these four layers exchanges of nutrients, gases, water, salts and waste products must take place. The barrier is efficient in

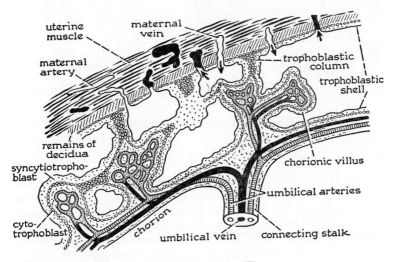

Fig. 15.7. Chorionic villi.

excluding most maternal parasites. But the spirochaete of syphilis can pass through and infect the foetus, so that it is born with 'congenital syphilis', and if a pregnant woman has German measles the virus often infects the foetus, producing gross abnormalities.

Transport by Placenta

The placenta behaves like a dialysing membrane in respect of diffusion of oxygen and carbon dioxide. Most substances with large molecules, including lipids and proteins, are transported by pinocytosis, and a considerable degree of selectivity is shown in this process. A large variety of antigens and antibodies pass through. Often this is beneficial in that the baby is born with some degree of immunity to most of the common infections. Tetanic infection of the umbilicus at birth, common in many parts of the world, is 90% lethal. But after building up the mother's antibody strength enough gets through the placental barrier to protect the newborn baby. But sometimes such transport is harmful, particularly in the case of Rhesus antigens and antibodies. Drugs may pass the barrier, so that the respiratory reflexes may be depressed or the genitalia affected by medication of the mother. The placenta is also permeable to hormones, and produces a considerable number of its own.

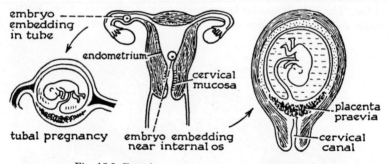

Fig. 15.8. Ectopic pregnancy and placenta praevia.

Ectopic Pregnancy. Fig. 8

Ordinarily the embryo embeds in the endometrium of the upper part of the body of the uterus, but it may embed elsewhere. If it is delayed it may not reach the uterus, but attempt development in the uterine tube, by far the commonest place for an ectopic pregnancy. Growth is normal for a time, but the tube becomes stretched and eventually the placenta separates, with serious haemorrhage requiring surgical intervention to save the mother's life, or the embryonic tissues may erode their way through the wall of the tube, again with serious haemorrhage into the abdominal cavity. At the time of removal the embryo is usually dead and surrounded by blood clot.

Abdominal Ectopic Pregnancy

If one tube is blocked, an ovum shed by the ovary on that side into the general peritoneal cavity of the abdomen may be fertilized there by spermatozoa ascending through the other tube. Indeed the ovum so fertilized may still find its way across the cavity and reach the uterus by the open tube and so establish a normal pregnancy. But the embryo may embed in the wall of the cavity or on one of the abdominal organs such as the intestine. Such abdominal ectopic pregnancies end as surgical emergencies with severe intraperitoneal haemorrhage.

Placenta Praevia. Fig. 8

If an embryo fails to embed before it reaches the cervical canal it is lost, for the cervical mucosa cannot serve for implantation. But the embryo may embed in the lower part of the body of the uterus, near the internal os. In this case the placenta will be formed over the os, as a placenta praevia. Growth is normal, but when the foetus comes to be born it must break through the placental substance before it can reach the cervical canal and vagina. The mother is in danger from haemorrhage, and the foetus from asphyxia.

Mature Placenta. Fig. 9

The mature placenta is disc-shaped, with a rough outer surface facing the uterine wall and a smooth inner surface facing the foetus. The rough surface is broken up into a number of cotyledons separated by

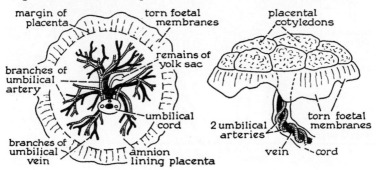

Fig. 15.9. Mature placenta.

shallow grooves. The smooth surface is covered by the amnion, which can be peeled off, and to this surface is attached the umbilical cord. The umbilical vein and the two umbilical arteries pass through the cord, twisting round each other. The direction of twisting is always the same and is probably dependent on molecular asymetry of the substances forming the cord, but the mode of twisting is not understood. On reaching the placenta the vessels branch on its inner surface. On expulsion from the uterus the remains of the amnion and chorion, the 'foetal membranes,' are found attached to the margin of the placenta.

Foetal and Maternal Plates. Fig. 10

The two surfaces of the placenta are made by the foetal and maternal plates. The foetal plate is made of the fused amnion and chorion, and in it ramify the umbilical vessels. The maternal plate consists of dense thickened trophoblastic tissue and a layer of decidua. When the placenta is expelled following childbirth it separates in this decidual layer. Part of the layer remains behind in the uterus and part forms the rough outer surface.

The Placental Septa. Fig. 10

The villi spring from the foetal plate and ramify between the plates. They are grouped into masses separated by the placental septa which dip

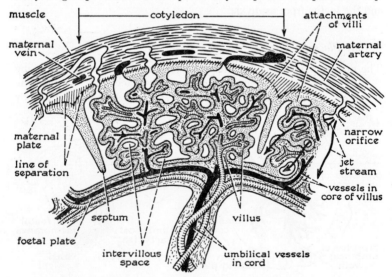

Fig. 15.10. Placenta and uterus (trophoblast dotted, decidua hatched).

between them. These septa are made of condensed trophoblastic tissue, and their attachments to the maternal plate make grooves on its outer surface, dividing it into cotyledons. The septa do not reach the foetal plate, so that the intervillous spaces of one cotyledon are continuous with those of the next.

The Villi and Intervillous Spaces. Fig. 10

The tips of the villi are attached to the maternal plate, and it is these attachments that hold the two plates together. The uterine arteries pour their blood into the intervillous space by rather narrow openings, and, if observations made on animals hold for man, the blood is delivered in a jet stream which takes it near the foetal plate. The villi are much branched and the branches joined to each other, so that the space is minutely subdivided and comparable in complexity to a capillary bed.

Binovular Twins. Fig. 11

Usually only one ovum matures at a time, the others being inhibited, probably by hormonal action. But two (or more) may mature together, either in the same ovary or in different ovaries, sometimes following the use of the recently developed fertility drugs. They may both embed and develop, each forming its own placenta and membranes. Where

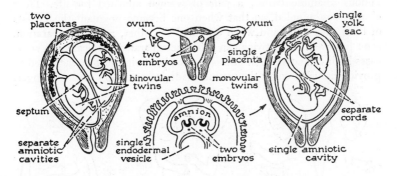

Fig. 15.11. Forms of twinning.

the membranes meet in the cavity of the uterus they form a septum between the two amniotic cavities. If the two placentae lie close together they may later fuse at their edges. Such twins are binovular or fraternal twins and may be of the same or different sexes.

Monovular Twins. Fig. 11

If at the 2-cell stage the two cells fall apart (probably on account of temporary asphyxia), each can develop into a separate embryo, giving eventually identical twins with the same chromosomal make-up. Formed in this way each of the twins will have its own placenta as in binovular twins. But division may occur later in development. Two embryos may form in a single embryonic plate and share a common amnion and endodermal vesicle. In this case there is but one placenta, with two umbilical cords attached to it. In the placenta there are likely to be extensive anastomoses between the umbilical circulations of the two embryos. This is one reason for tying the cord in two places and cutting between the ligatures when the baby is born, for an unborn twin, whose presence was unsuspected, might bleed to death through its brother's cord. Twinning is the commonest gross anomaly of development, occurring once in about 100 live births. Monovular twins are only one third as common as the binovular kind.

Conjoined Twins. Fig. 12

Where twins share a single amniotic cavity they may only partially separate or, though originally separate, they may come into mutual contact and fuse. In this way various kinds of double monsters, all rarities, may be formed, often very symmetrically. The fusion may be superficial, involving only skin and subcutaneous tissues, as in the famous 'Siamese twins', a pair of women attached for life. Their circulations mingled, so that when one became pregnant the breasts of both gave milk, and when one died the other, refusing separation, soon followed. Such twins can often be separated surgically. In some cases both can survive, in others the weaker twin must be sacrificed to provide tissue to cover the wound made in separation of the stronger.

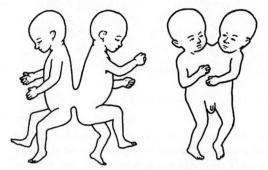

Fig. 15.12. Conjoined twins.

Partial Subdivision. Figs. 12 & 13

Partial subdivision gives another series of double monsters. Two-headed human babies die, probably because the two guts fail to join or the urinary apparatus is malformed. Two-headed snakes have been known to grow up, and there seems no absolute reason why a human monster should not do so. There is, in fact, a recent report from Pakistan of a dicephalus tripus quadribrachius, as in fig. 13, which appears healthy at the age of two, with shared anal and urogenital orifices. When the heads of two embryos are joined face to face the maxillary and mandibular processes of the one may react with those of the other. This gives a peculiar symmetrical monster with one face looking to one side and another face looking to the other, each face including contributions from both embryos.

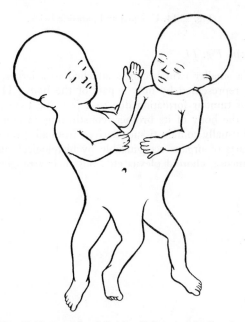

Fig. 15.13. Partial subdivision living conjoined twins.

Unequal Twins. Fig. 14

When twins share a placenta there are often extensive anastomoses between their circulations. If one of the twins is a little more advanced than the other it is likely to get more than its fair share of blood, so that as time goes on the differences of development become more marked. Eventually the smaller twin is often reduced to a formless mass of tissue, expelled with the normal twin when it is born.

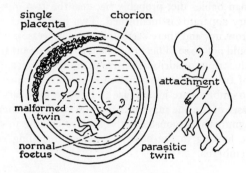

Fig. 15.14. Unequal and parasitic twins.

Parasitic Twin. Fig. 14

Sometimes a smaller twin becomes attached to a larger, the smaller often being represented by only a part of the body. This gives the possibility of tumour formation due to a parasitic twin more or less included in the body of its brother, constituting an embryoma. The tumour will usually contain tissues derived from all three germ layers, including parts of organs recognizable by histological study, such as intestinal mucosa, choroid plexus etc., but their arrangement will be chaotic.

16

The Abdomen

Greater Omentum and Spleen. Fig. 1

At 7 mm. the stomach has already rotated so that the greater curvature looks to the left and the lesser curvature to the right. A small lesser sac lies behind the stomach, between it and the mesogastrium, the part of the mesentery that attaches the greater curvature to the body wall. Later the mesogastrium expands so as to hang from the greater curvature as a large double sheet, the greater omentum, with the lesser sac extending between its layers. The spleen appears as a thickening on the left face of this sheet, opposite the upper end of the stomach. It comes to bulge to the left and its mesenchyme is filled with lymphoid tissue.

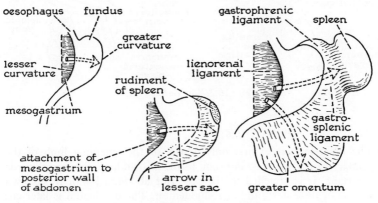

Fig. 16.1. The greater omentum.

Gastrophrenic and Lienorenal Ligaments. Fig. 1

The mesogastrium is divided by the splenic attachment into several parts, still continuous with each other. The uppermost is the gastrophrenic ligament. Between the spleen and the stomach is the gastrosplenic and between the spleen and the posterior abdominal wall the

159

lienorenal ligament. The rest of the mesogastrium forms the greater omentum itself. At this stage the dorsal attachments of the gastro-phrenic and lienorenal ligaments and of the greater omentum are to the midline of the abdomen, along the course of the aorta.

The Portal Vein. Figs. 2 and 7.12

Parts of both vitelline veins are preserved, with their cross-anastomosis dorsal to the gut, to form a single channel, the early portal vein, leading blood coming from the gut by the superior mesenteric vein to the liver. The unwanted portions of the vitelline veins disappear. As the spleen develops in the mesogastrium a new vein, the splenic, is formed to drain it into the portal system. The inferior mesenteric vein is formed later from a capillary plexus bringing blood from the caudal part of the gut region.

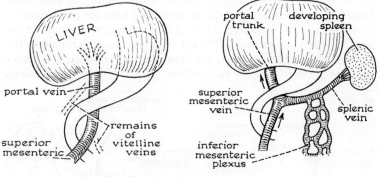

Fig. 16.2. The portal vein.

Stomach and Duodenum. Fig. 3

The duodenum is at first only a part of the emerging limb of the primary loop of the gut, but soon forms its own loop, directed towards the right. It is attached to the posterior abdominal wall by its own section of mesentery, the mesoduodenum. The duodenum is later cut off from the stomach by a thickening of the circular muscle of the gut, the pylorus.

The Pancreas. Fig. 3

The larger dorsal and smaller ventral pancreatic buds lie close together and both grow out into the mesentery. The two buds fuse and their ducts anastomose. The dorsal pancreas makes the upper part of the head and the neck, body and tail of the adult pancreas, while the ventral bud makes only the lower part of the head and the uncinate process. Most of the secretion of the dorsal pancreas passes through the anastomosis to reach the duodenum by the duct of the ventral pancreas, which becomes the main pancreatic duct of the adult. The duct of the dorsal pancreas persists as the accessory duct, usually very small.

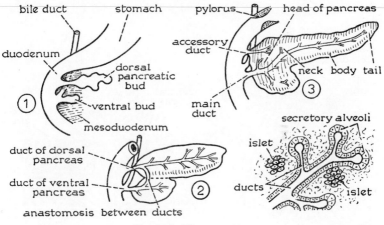

Fig. 16.3. The pancreas.

Pancreatic Islets. Fig. 3

The ducts and acini of the pancreas develop like those of a salivary gland. But some of the terminal buds separate from the duct system and become islets, concerned with the formation of insulin. Thus the pancreas, which has both external and internal secretions, resembles both the salivary and the parathyroid glands in its development. Insulin formation is active before birth, and some of the hormone may reach the mother from the foetus through the placenta.

Biliary Units of the Liver. Fig. 4

The early liver was made up of liver trabeculae, at first solid, surrounded by sinusoidal blood spaces derived from the endothelium of the vitelline veins, all embedded in a framework of septum transversum mesenchyme. The trabeculae branch repeatedly to form the bile duct system, and later hollow out to give each duct its lumen. A set of radially arranged liver plates spreads from the duct. The plates are perforated and are only one cell thick, so that each cell is exposed to the sinusoids on at least two surfaces. The duct and its plates form a biliary unit,

G

Blood in the Liver. Fig. 4

Blood vessels, derived from the sinusoids, develop in relation to the biliary units. Near the bile duct, in the centre of the unit, is a branch (distributory) of the portal vein. Between the units are efferent veins leading the blood to the developing vena cava and so to the heart. Thus the blood brought to the liver by the portal vein has to pass through the sinusoids between the plates, and is then collected by the efferent veins.

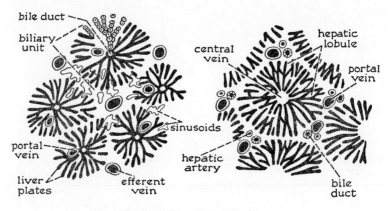

Fig. 16.4. Structure of the liver.

Liver Lobules. Fig. 4

Eventually the plates are re-arranged round the efferent veins to form hepatic lobules of the adult type. The efferent vein becomes the central vein of the lobule, which includes plates from several adjacent biliary units. This leaves the bile duct, with its accompanying distributory of the portal vein, in the tissue between the lobules. Later branches of the hepatic artery spread along the ducts, giving, with the neighbouring connective tissue, the characteristic Glisson's capsule and its contents in the adult.

Openings of the Coelom. Fig. 5

With the folding of the embryo the two openings from the intra-embryonic coelom into the extra-embryonic coelom come to lie on the ventral surface of the body. They are found caudal to the liver, on either side of the primary loop of the intestine and its mesentery. The emerging limb of the loop is attached to the liver by a mass of mesenchyme,

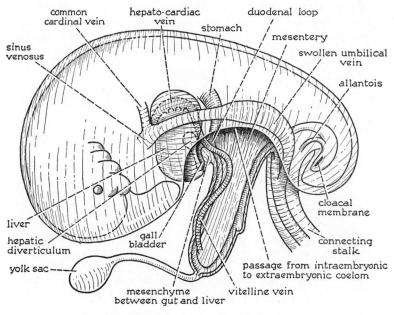

Fig. 16.5. The embryo seen through the amnion.

originally a part of the septum transversum. In this mass runs the stalk of the hepatic diverticulum which becomes the bile duct, the developing portal vein, and, in later development, the hepatic artery.

The Lesser Omentum. Fig. 6

The stomach and duodenum are withdrawn from the liver, stretching the mesenchyme that attached them to form the lesser omentum. At its free margin, where the bile duct, portal vein and hepatic artery pass to and from the liver, the lesser omentum remains thick, but elsewhere it becomes very thin. The thick margin was originally level with the anterior abdominal wall, but is gradually drawn back into the abdomen. So the right and left parts of the intra-embryonic coelom come to communicate across it giving a single peritoneal cavity.

Coronary Ligament. Fig. 7

In the early embryo the heart is so large that it occupies the whole width of the body. It is separated from the liver by the pericardial cavity and by a layer of septum transversum mesenchyme. A condensation of this mesenchyme forms the central tendon of the diaphragm. Between this tendon and the liver the mesenchyme forms the coronary ligament, which attaches the bare area of the liver to the diaphragm. In the adult, then, the septum transversum is represented by:

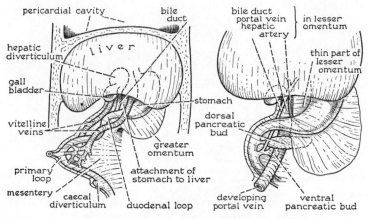

Fig. 16.6. Lesser omentum and its contents.

(1) the central tendon, (2) the coronary ligament, (3) the stroma of the liver, including Glisson's capsule and (4) the lesser omentum.

Domes of the Diaphragm. Fig. 7

The lungs and pleurae spread round the sides of the heart. In doing so they split the thoracic part of the body wall into two layers, one lateral to the lung, the thoracic wall, the other medial to the lung, the fibrous pericardium. The thorax is widened in the process. At the same time the liver is growing rapidly and so widening the abdomen. Thus a part

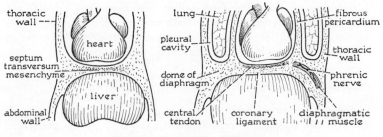

Fig. 16.7. Parts of the diaphragm.

of the body wall is drawn in on each side, between the pleural and peritoneal cavities, to form the connective tissue stroma of the domes of the diaphragm.

Muscle of Diaphragm. Fig. 7

Muscle-forming tissue, derived from the somites, migrates into the domes, drawing its nerve supply with it. At this stage the diaphragm still lies in the cervical region and the muscle is derived from cervical somites. So in the adult the muscle is supplied by the phrenic nerve, which has cervical roots.

The Pleuroperitoneal Membranes. Fig. 8

The mesonephros is originally formed in the thoracic region. Later new glomeruli and tubules develop at its caudal end, while its cranial end atrophies to give a residual ridge. This ridge comes to lie dorsal to the liver, and here it forms the pleuroperitoneal membrane. The membrane is placed vertically beside the pleuroperitoneal opening, with its free margin directed caudally.

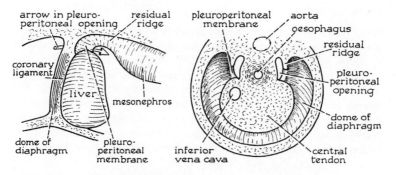

Fig. 16.8. The pleuro-peritoneal membrane.

Closure of Pleuroperitoneal Openings. Figs. 8 & 9

The pleuroperitoneal membranes swing up into a horizontal position, so that their free edges come into contact and fuse with the dense mesenchymal tissue surrounding the oesophagus. Thus the openings are closed and the diaphragm completed. Muscle from the domes spreads into the membranes. The adult diaphragm, then, is made up of septum transversum (central tendon), the tissue round the oesophagus, the body wall (domes), the residual mesonephric ridges (pleuroperitoneal membranes) and muscle from cervical myotomes. The weakest part is the last formed, the parts contributed by the membranes, and here diaphragmatic hernia may occur later, more often on the left side, but sometimes bilateral.

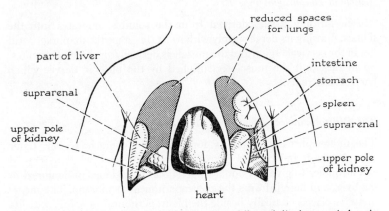

Fig. 16.9. Viscera in pleural cavities in severe bilateral diaphragmatic hernia.

Coeliac Artery. Fig. 10

At its first appearance the coeliac artery is a short trunk passing to the greater curvature of the stomach in the mesogastrium. When the spleen develops in the mesogastrium it takes blood from this artery, the terminal twigs now forming the short gastric and left gastro-epiploic, while the main trunk becomes, for the greater part of its length, the splenic artery. As the diaphragm develops a new vessel, the left gastric, passes directly to the lesser curvature of the stomach by 1 shorter route across its left crus. The hepatic artery begins as a small branch of the coeliac to the pyloric end of the stomach and the duodenum, the gastro-duodenal artery of the adult. Later it spreads along the bile duct to reach the liver and gall bladder. The left gastric and hepatic, hooking round the lesser sac, constrict it whereas the splenic artery, lying entirely behind the sac, does not.

The Early Intestine. Fig. 11

The stomach projects to the left, the duodenum to the right, and a sharp duodeno-jejunal flexure to the left again. The rest of the intestine has three straight limbs of about equal length, the emerging limb, between the flexure and the yolk stalk, the returning limb between the yolk stalk and the splenic flexure, and the terminal limb. The caecal diverticulum on the returning limb marks the union of the small and large intestines. The greater part of the emerging and returning limbs lie in the extra-embryonic coelom. At this stage the yolk stalk shrinks and breaks, leaving the intestine free to coil.

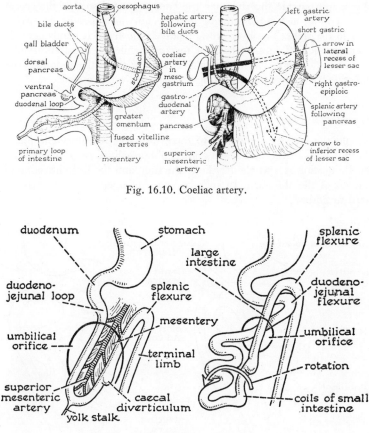

Fig. 16.10. Coeliac artery.

Fig. 16.11. Initiation of rotation.

The Duodeno-jejunal Flexure. Fig. 11

The flexure moves to the left, insinuating itself between the returning and terminal limbs. This is the first move in the rotation of the gut. The abdominal cavity is relatively small and mostly occupied by the liver, so there is no space for coiling there, and the umbilical orifice has narrowed so that there is only room for the two limbs to pass through. But the extra-embryonic coelom is still capacious and here the rapidly elongating intestine is thrown into a knot of coils.

Rotation of the Gut. Fig. 12

While still outside the abdomen the knot is rotated so that the caecal diverticulum, now differentiating into caecum and appendix, is carried over to the right. The liver grows relatively slowly as it hands over its haemopoetic activities to the bone marrow, while the extra-embryonic coelom is closing in. So, with relatively more space in the abdomen, the intestine withdraws into it, taking up the adult arrangement.

Malrotation. Fig. 12

Should the intestinal knot fail to rotate in its extra-embryonic phase the gut will, in the adult, show gross abnormalities of position. The small intestine as a whole may lie to the right and the large intestine to the left of the abdomen. In such cases an inflamed appendix may be hard to find.

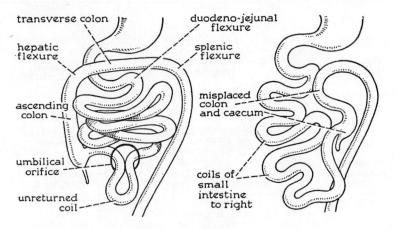

Fig. 16.12. Completion of rotation and malrotation.

The Appendix. Fig. 12

The caecal diverticulum is at first blunt ended. Later its terminal part is drawn out as a relatively narrow appendix.

The Mesenteries. Figs. 12 & 13

As the gut rearranges itself in the abdominal cavity several sections, with their mesenteries, become attached to the posterior abdominal wall or other organs. A relatively firm attachment of the colon to the duodenum by a duodenocolic ligament fixes the hepatic flexure. Later the duodenum and mesoduodenum and the ascending and descending colons, with their sections of the mesocolon, are also attached. So the originally continuous general mesentery of the gut is represented in the adult by a number of separate sections, the main mesentery of the jejunum and ileum attached along the line of the superior mesenteric artery, the transverse mesocolon and the pelvic mesocolon. Most of the mesogastrium is preserved in the greater omentum, but a part is attached so that the lienorenal ligament no longer reaches the midline. The attachment of the mesogastrium and mesoduodenum to the posterior abdominal wall also brings about the attachment of the pancreas, which has been developing in them.

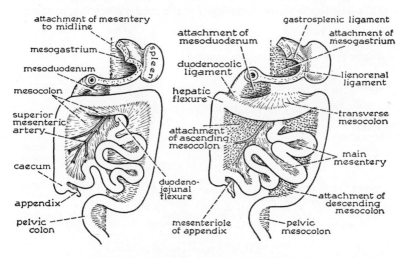

Fig. 16.13. Adhesions of the mesentery (dotted).

Atresia and Stenosis of the Intestine

Any part of the intestine may be blocked (atresia) or narrowed (stenosis). The mechanism is not known, but it has been suggested that the rapid elongation of the gut at the time of formation of the primary and secondary loops may stretch the epithelium so that the lumen is narrowed or its continuity broken. Possibly too this is a period of

difficulty in blood supply leading to tissue death. So one or more segments of the intestine may have a deficient lumen or be altogether
missing, with their associated segments of mesentery. For its length the
duodenum is the part most often affected. The bile duct may also be
closed.

Pyloric Stenosis

Narrowing of the pyloric canal with marked hypertrophy of the pyloric
sphincter is common but does not appear till a few weeks after birth.
Its mechanism is entirely unknown. The condition is operable by simple
incision of the muscle without cutting the mucous membrane.

The Transverse Mesocolon. Fig. 14

After the rotation of the gut the transverse mesocolon and its membranes come to lie across the abdomen dorsal to the greater omentum.
They now fuse with the dorsal sheet of the greater omentum, so that in

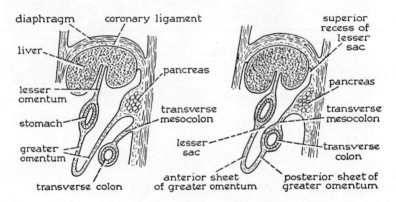

Fig. 16.14. The transverse mesocolon.

the adult the sheet appears to be attached to the colon, and to be
continued into the mesocolon. This gives the lesser sac its adult posterior boundaries: the posterior sheet of the greater omentum, transverse colon, transverse mesocolon, pancreas and posterior abdominal
wall. Failure of attachment of a mesentery or omentum is common,
giving extra mobility in the region concerned. Excessive attachment is
usually due to inflammation rather than embryonic abnormalities.

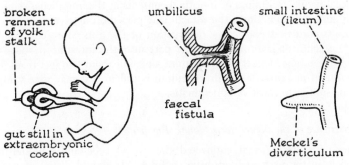

Fig. 16.15. Faecal fistula and Meckel's diverticulum.

Meckel's Diverticulum. Fig. 15

Before the return of the gut to the abdomen the yolk stalk is thinned and broken. If the stalk persists as an open tube till birth it will be cut with the umbilical cord and come to open at the umbilicus. This gives a

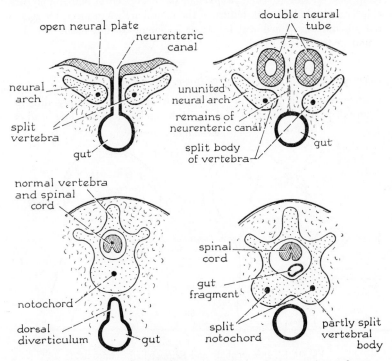

Fig. 16.16. Anomalies ascribed to persistence of neurenteric canal.

'faecal fistula', with loss of intestinal contents at the umbilicus, requiring surgery. Sometimes the attachment persists, but is reduced to a fibrous cord that may cause obstruction of the gut. More commonly the attachment is broken, but the part near the intestine persists as a blind-ending Meckel's diverticulum, opening about two feet from the ileocaecal junction. If the diverticulum is inflamed it may give symptoms similar to those of appendicitis.

Persistence of the Neurenteric Canal. Fig. 16

The neurenteric canal connected the neural tube and endodermal vesicle in the very early embryo (fig. 3.4.). It should dissapear completely, but sometimes persists as a hollow duct or a fibrous cord, either isolated or attached to the gut or spinal cord, and the anomalies produced by it may affect any level of the trunk, for the primitive node through which it passes moves posteriorly as successive somites are added. Its presence in the median plane may interfere with neural fold formation so that the neural tube fails to close or closes as two tubes lying side by side (diastematomyelia), or the vertebral bodies may be split (vertebral cleft) with failure of the neural arches to meet. Or again the gut may show a dorsal fistula or fragments of gut tissue may be found in the spinal canal.

Iniencephaly and Klippel-Feil Syndrome. Fig. 17

Iniencephaly is a profound lethal disturbance also possibly due to failure of proper separation of neural tube and endoderm. The head is

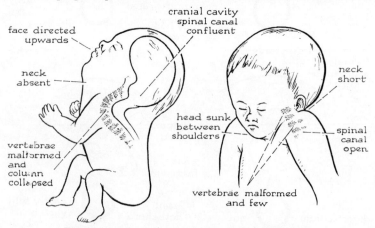

Fig. 16.17. Iniencephaly and Klippel-Feil Syndrome.

joined directly to the trunk with no neck, giving a humpty-dumpty appearance and the face looks upwards rather than forwards. The vertebrae are malformed, open posteriorly and sometimes anteriorly and the spine grossly distorted and the viscera show multiple anomalies. The Klippel-Feil syndrome is a milder but probably related defect. The neck is short and the vertebrae divided so that the infant cannot move its head, but may survive.

The Suprarenal, Kidney and Cloaca

Suprarenal Gland. Fig. 1

The suprarenal develops from two distinct sources that make separate organs in fishes, but are combined in mammals. The medulla is derived from cells of the neural crest that first follow the cells migrating to form the sympathetic ganglia, but later continue their migration round the sides of the aorta. The cortex comes from a thickening of the coelomic epithelium between the mesentery and the mesonephros. At first the two rudiments lie side by side, but later the medulla makes its way inside the cortex. The suprarenal is well developed in foetal life and appears to be physiologically active. The bulk of its substance is made by anastomosing cords of provisional cortex. The permanent cortex remains relatively undeveloped till early infancy, when the provisional cortex disappears.

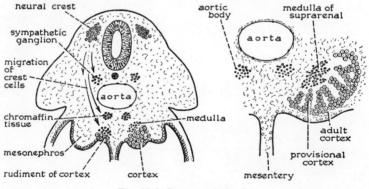

Fig. 17.1. Suprarenal gland.

Aortic Bodies. Fig. 1

Small masses of suprarenal tissue, particularly of medullary type (chromaffin tissue), are frequently found near the aorta. A pair of larger

174

masses, the aortic bodies, are conspicuous at birth but become obscure later.

Calyces and Collecting Tubules. Fig. 2

The free end of the metanephric bud expands to form the renal pelvis,

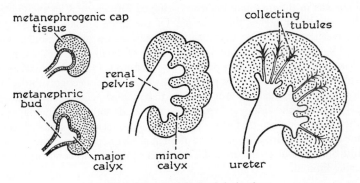

Fig. 17.2. Metanephric bud derivatives.

while its stalk elongates to form the ureter. From the pelvis new outgrowths arise, the major and minor calyces, and from these, in turn, grow out the collecting tubules and their branches.

Renal Tubules. Fig. 3

The tissue of the metanephrogenic cap becomes lobulated and rearranged so as to form hollow vesicles, the renal tubules, opposite the blind, swollen ends of the collecting tubules. Each renal tubule becomes S-shaped. At one end a glomerulus and Bowman's capsule develops

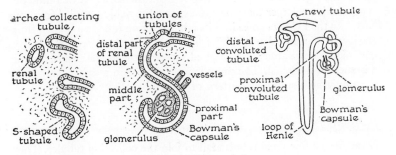

Fig. 17.3. Kidney tubules.

while the other end joins the collecting tubule. The structure of the renal tubule at this stage resembles that of a fully developed mesonephric tubule.

Loop of Henle. Fig. 3

Whereas in the mesonephros the S-shaped form is final, in the kidney the tubule undergoes modification. Of the three limbs of the S, that nearest the glomerulus becomes the proximal convoluted tubule. The middle limb is drawn out into the loop of Henle, forming both its descending and ascending parts. The third limb forms the distal convoluted tubule. New buds grow from the earlier collecting tubules, to form, with residual nephrogenic cap tissue, new generations of excretory units as the kidney grows. These new buds form a distinct layer in the most superficial part of the cortex, which thus acts as a growth zone for the foetal kidney. By the time of birth all the nephrons are developed but the loops of Henle are still short and the power to concentrate the urine limited.

Migration of Kidney. Fig. 4

When they are first formed the kidneys are very small and lie in the pelvic cavity, caudal to the bifurcation of the aorta. Later they migrate cranially, passing out of the pelvis onto the posterior abdominal wall and so to their adult positions just below the suprarenals. As they rise they rotate so that the renal pelvis, which was originally ventral, comes to lie on the medial margin, and they become strongly lobed.

Blood Supply of the Kidney. Fig. 4

While still in the pelvis the kidney is supplied from the iliac arteries, but as it rises it gets new supplies directly from the aorta. At this time the suprarenals are larger than the kidneys and are richly supplied from the aorta. As the kidney comes into its adult position it takes a branch from the lowest suprarenal artery. This branch eventually enlarges to become part of the renal artery, the original vessel to the suprarenal becoming a branch of the renal. The phrenic artery arises in a similar way from the uppermost suprarenal artery of the embryo.

Polycystic Disease

Sometimes a few renal tubules fail to join the collecting tubules. They become filled with secretion, forming cysts in the kidney substance. An excessive number of cysts (polycystic disease) severely restricts the efficiency of the kidney.

Abnormal Renal Artery

Usually, with the development of the renal artery, the temporary branches from the aorta that supplied the kidney during its migration are lost. One or more may persist and compress the ureter, leading later to dilatation and infection of the renal pelvis. Again such an abnormal artery may become atheromatous, so that the blood supply to the section of kidney it supplies is cut off. The juxtaglomerular cells of the afferent arterioles secrete renin, resulting in general hypertension. The anomaly can be corrected surgically.

Horse-shoe Kidney. Fig. 4

The kidney may be arrested anywhere in its ascent. In this case it may develop and function normally, but is more liable to disease than properly placed kidneys. As the two kidneys cross the common iliac arteries in their ascent they are forced together and may fuse, forming a horseshoe kidney. This combined structure continues its ascent till it comes up against the inferior mesenteric artery. The fusion interferes with rotation, so that the ureters are usually found emerging from the ventral surface of the horseshoe. These kidneys are particularly liable to disease.

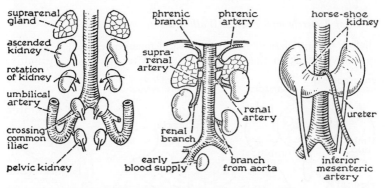

Fig. 17.4. Migration of kidney and horseshoe kidney.

Cloaca. Fig. 5

The cloaca is the dilated end of the hind gut, caudal to the attachment of the allantois. It receives, nearer its ventral than its dorsal surface, the two mesonephric ducts into which the metanephric buds open. It is separated from the amniotic cavity by the cloacal membrane.

Cloacal (Uro-rectal) Septum. Fig. 5

The mesenchymal tissue between the gut and the attachment of the allantois, assisted by mesenchyme lying lateral to the cloaca, proliferates (crosses in the diagram). It forms the cloacal septum, which spreads caudally, cutting the cloaca into a rectum dorsally and a urogenital sinus ventrally. The gut ends in the rectum, while the allantois and mesonephric ducts end in the urogenital sinus. Eventually the septum meets and fuses with the deep surface of the cloacal membrane.

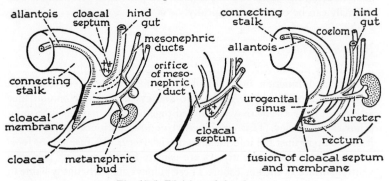

Fig. 17.5. Division of the cloaca.

Genital Tubercle and Ventral Wall of the Abdomen. Fig. 6

Before folding of the embryo the cloacal membrane reached from the primitive streak to the connecting stalk. After folding the primitive

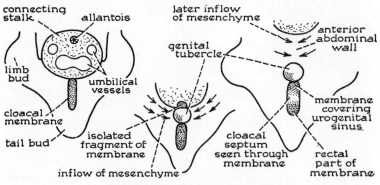

Fig. 17.6. Anterior abdominal wall.

streak region forms the ventral surface of the tail bud and, eventually, the anococcygeal region, while the membrane still reaches the attachment of the stalk. Mesenchyme flows in from either side to cut off a small fragment of the membrane, which remains attached to the stalk, from the main portion. The first inflow builds the genital tubercle, and a later inflow the part of the ventral wall of the abdomen that lies between the tubercle and the stalk.

Separation of the Vas (Ductus) Deferens and Ureter. Fig. 7

The cranial part of the urogenital sinus widens to form the bladder, into whose apex the allantois opens. The rest of the sinus forms the urethra, the whole in the female, only a part in the male. The mesonephric duct forms the vas deferens and the stalk of the metanephric bud, the ureter, but these two channels still open by a common stem into the

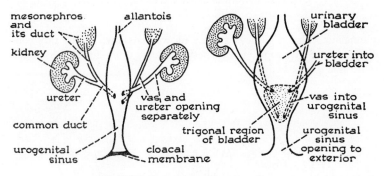

Fig. 17.7. Separation of ureter and vas.

bladder. The common duct is taken into the bladder wall, making the trigonal region, very much as the pulmonary vein is taken into the left auricle, making the smooth part of its wall. So the two ducts come to open separately, and semen and urine can be kept apart in the adult.

Retrocaval Ureter. Fig. 8

The ureter may follow an unusual path in its ascent, passing between the aorta and inferior vena cava. This may give no trouble but sometimes the ureter is compressed by the vein and its upper part becomes dilated from back pressure. This anomaly is sometimes ascribed to a peculiar development of the inferior vena cava in the embryo, but it seems more likely that the vein is developed normally but the ascent is

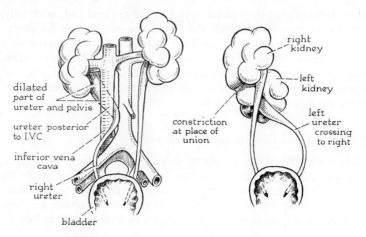

Fig. 17.8. Retrocaval ureter and crossed dystopia.

anomalous. That a kidney can lose its way is shown in the rare condition crossed dystopia, where both kidneys, often fused, lie on the same side of the body.

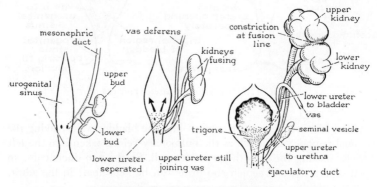

Fig. 17.9. Doubled metanephric bud.

Duplicated Kidney. Fig. 9

Duplicated kidney is relatively common. Two metanephric buds develop from the mesonephric duct, one above the other. When the lower end of the mesonephric duct is taken into the wall of the urogenital sinus the ureter of the lower kidney reaches the sinus first and achieves the normal position, opening on the upper lateral corner of the trigone.

The ureter of the upper kidney reaches the sinus late and opens nearer the ejaculatory duct, usually into the prostatic urethra. It is then the upper kidney, or upper segment of the fused pair, that requires removal.

Phallus and Destruction of Cloacal Membrane. Fig. 10

The genital tubercle grows ventrally, forming the phallus, which will develop later into the penis or clitoris. As it grows it draws out a diverticulum of the urogenital sinus, including a part of the cloacal membrane.

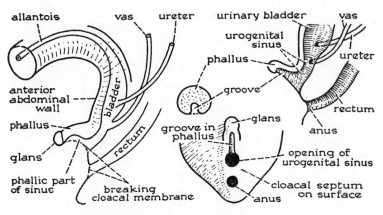

Fig. 17.10. Anus and phallus.

At this stage the membrane is thin, so that the free edge of the cloacal septum can be seen through it. It breaks down altogether. The rectum opens directly to the exterior at the anus, for there is, as yet, no anal canal. The urogenital sinus too opens directly, but is also carried forward as a groove on the underside of the phallus. The terminal, solid part of the phallus forms its glans.

Anal Canal. Fig. 11

The perineal musculature develops round the cloaca. At first it surrounds both the anal and urogenital regions, forming a generalized sphincter cloacae, but later a fibrous perineal body separates the anal sphincter from the urogenital muscles. The sphincter raises a circumanal rim on the surface and a new section of gut, the anal canal, is developed, its epithelium probably derived partly from that of the rectum and partly from that of the skin. The first motion, of meconium,

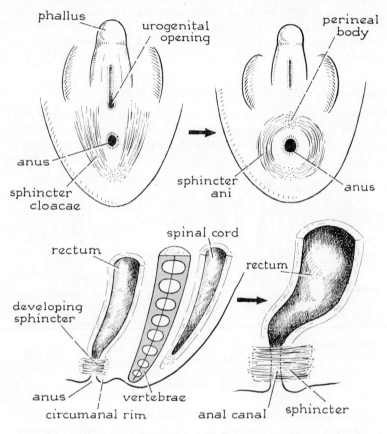

Fig. 17.11. Anal canal.

a black substance consisting chiefly of dead intestinal epithelial cells, as usually passed after birth.

Imperforate Anus and Rectal Fistula. Fig. 12

At birth there may be a membrane closing the anus and soon bulged by meconium which has to be broken artificially. Or the whole anal region may be plugged by a solid mass of epithelial cells. Such barriers have been ascribed to persistence of the cloacal membrane, but this is a relatively weak structure. They are more likely to be due to a proliferation of epithelial cells blocking a previously existent canal, as is known to occur in pyloric stenosis.

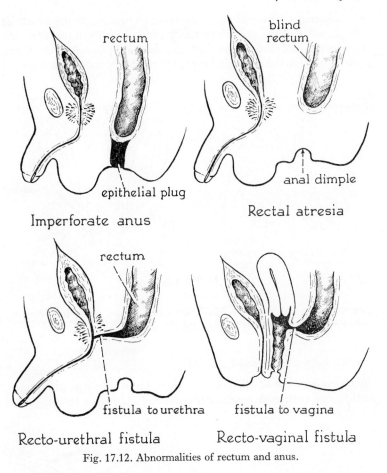

Fig. 17.12. Abnormalities of rectum and anus.

The terminal part of the rectum may be lost so that it ends far from the surface, sometimes with the anal canal represented by a blind dimple. When the cloacal septum fails to develop the rectum and urogenital sinus derivations remain united, the rectum usually opening into the upper urethra in the male and the vagina in the female. The girl can live as she can pass faeces; the boy must die unless rescued by surgery.

The Müllerian Ducts. Fig. 13

The Müllerian duct (paramesonephric duct) arises in both sexes as a groove in the mesonephric ridge, lateral to the mesonephric duct. The groove becomes covered over by fusion of its lips, but in the female its cranial end remains open into the peritoneal cavity throughout life The free caudal end of the duct crosses the mesonephric duct, and makes its way to the urogenital sinus. On the dorsal wall of the sinus the

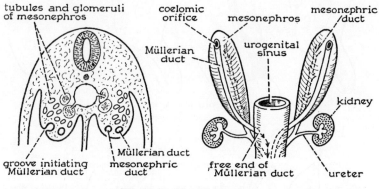

Fig. 17.13. The female ducts.

two ducts fuse to form the Müllerian cord, at first solid, but later developing a lumen which opens into the sinus.

Uterus, Vagina and Hymen. Fig. 14

At the attachment of the cord the sinus forms an upgrowth, which, with a mesenchymal condensation around it, makes the vagina. A

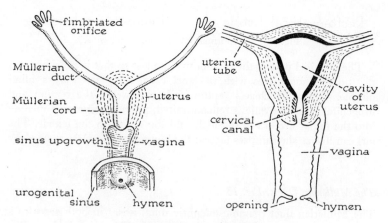

Fig. 17.14. Uterus and vagina.

more massive condensation round the cord makes the uterus. The vaginal lumen expands while its opening into the sinus remains narrow, so giving the hymen.

Opening Out the Sinus. Fig. 14

When the hymen is first formed the vagina opens into the sinus about half way along its length. Later the caudal part of the sinus is opened out so as to form the vulvar vestibule. This brings the hymen and vaginal opening to the surface of the body. The vulva, where it is formed of surfaces derived from the cloacal septum and urogenital sinus, must be lined with endoderm. But there is no histological distinction between this region and the surrounding skin in the foetus, nor can the extent of the endoderm be marked out in the adult.

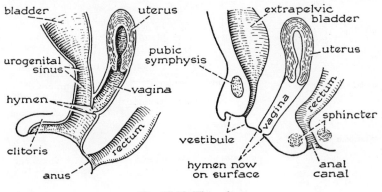

Fig. 17.15. The vulva.

Pelvic Viscera of the Newborn. Fig. 15

At birth the pelvis is very small, so that the whole of the bladder and a part of the uterus lie outside it, in the main abdominal cavity. An anal sphincter has developed round the terminal part of the gut, which is bent sharply on the rectum to form the anal canal. Here again the original position of the cloacal membrane, that marks the junction of ectoderm and endoderm, can no longer be distinguished.

Imperforate hymen, due to overgrowth of tissue at the opening of the vaginal upgrowth into the urogenital sinus, usually gives no trouble before the onset of menstruation, when the accumulation of menses makes a painful swelling. Absence of the hymen due to excessive expansion of the vaginal opening may cause difficulties in countries where its presence is valued, but it can be simulated by plastic surgery.

Umbilical Region. Fig. 16

When the embryo is first folded off the umbilical region, that is, the region enclosed in the attachment of the amnion, is very extensive. It encloses a part of the septum transversum, into which the hepatic diverticulum and trabeculae are growing, the yolk stalk with the openings of the coelom on either side, and the connecting stalk (compare fig. 16.5.). The region remains large while the gut hangs outside

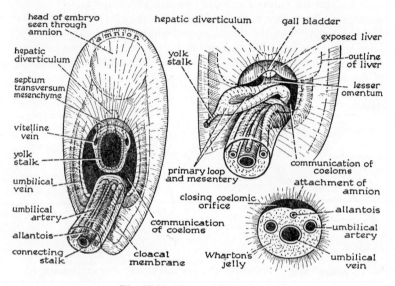

Fig. 17.16. The umbilical region.

the embryo into the extra-embryonic coelom. But eventually the attachment of the amnion to the embryo closes in and reduces the umbilical region. The yolk stalk breaks, the opening of the coelom, now single, is closed and the abdominal wall covers all but a small area. The umbilicus itself is better preserved in the human adult than in other mammals. With the loss of hair and adoption of the upright posture, it has become an ornament to the abdomen.

Attachments of Liver. Fig. 17

The early liver projects equally into the abdomen on the left and right, and the gall bladder lies in the median plane. The greater part of the cranial surface is attached to the developing diaphragm by the mesenchyme of the coronary ligament. The left umbilical vein, after skirting the left border of the umbilical orifice in the body wall, enters the substance of the liver to join the ductus venosus.

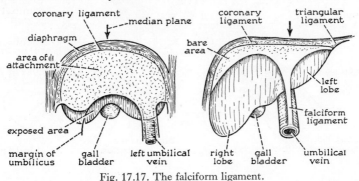

Fig. 17.17. The falciform ligament.

Falciform Ligament. Fig. 17

The liver deepens on the right and thins on the left. The coelom spreads over the greater part of its ventral surface, so as to cut down the bare area. The umbilical vein is drawn out of the body wall so as to lie in the free edge of the falciform ligament. The ligament is made of the substance of the wall (not from a 'ventral mesentery', as sometimes stated). The liver as a whole shifts to the right as the other organs of the abdomen develop, so that the umbilical vein and falciform ligament come to lie in the median plane, and the gall bladder to the right.

Ectopia of the Bladder. Fig. 18

The full extent of the cloacal membrane was from the attachment of the connecting stalk, eventually the umbilicus, to the anal region. Normally only the perineal part persists to the time it breaks down to make openings for the urogenital sinus and rectum, while the abdominal part is invaded by mesenchyme (fig. 6). If, however, the whole of the membrane persists it may break down later throughout its length. Then the anterior wall of the abdomen below the umbilicus is split, and the bladder wall, with the ureteric openings, is exposed on the body surface. If the split involves the genital tubercle two separate half-penises may result. Less extensive splitting gives epispadias, that is, opening of the urethra on the dorsum of the penis.

Bicornuate Uterus. Fig. 18

Failure of proper fusion of the Müllerian ducts may give a uterus with two separate horns, the normal state in animals which bear many young

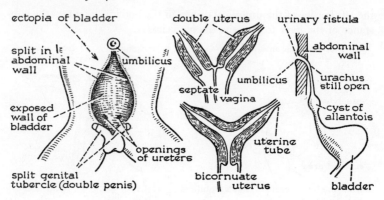

Fig. 17.18. Urogenital anomalies.

at a birth but leading, in pregnant women, to a transverse lie of the foetus and consequent difficulty in delivery. Or the uterus may be completely divided into two separate organs, or it may have a normal external shape but be divided internally by a septum. The vagina again may be septate, a condition which can usually be corrected surgically.

Urinary Fistula. Fig. 18

Normally the allantois, which stretches from the umbilicus to the apex of the bladder, is reduced, long before birth, to a solid fibrous cord, the urachus. Occasionally it may remain open locally and dilate to form a cyst. If it stays open throughout its length urine may escape, after birth, from the cut cord or umbilicus, giving a urinary fistula.

Umbilical Anomalies

At birth, then, there may be escape of urine from the umbilicus from an open urachus, or of faeces from a persistent yolk stalk. The gut may protrude into the cord if the extra-embryonic coelom has not been obliterated, giving umbilical hernia, or it may be connected to the umbilicus by a fibrous cord, the remains of the yolk stalk. The rectus muscles may fail to come into their proper positions on either side of the umbilicus, giving a wide membranous linea alba and a spread umbilicus. All these anomalies are of surgical importance.

18

The Genital Organs

The Gonad. Fig. 1

The gonad appears as a swelling on the medial surface of the meso-nephros, so that it is, from the first, close to the mesonephric tubules and glomeruli. The mesonephric duct lies nearby and parallel to it, and the female duct lateral to the mesonephric duct. The gonad is relatively short, so that the mesonephros extends beyond it at both ends.

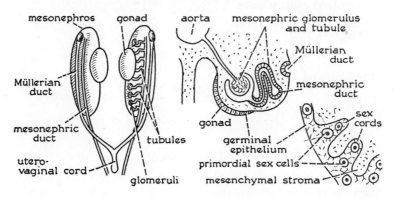

Fig. 18.1. Early gonad.

Primordial Germ Cells. Fig. 1

The coelomic epithelium covering the gonad is thickened to form the germinal epithelium, but there is no clear boundary between this and the underlying dense mesenchyme. Primordial germ cells from the endoderm of yolk sac have earlier migrated into this region. These cells are larger than those of germinal epithelium or mesenchyme, and their shape is rounded. They get embedded in cords of cells, the sex cords, which develop in the dense mesenchyme of the gonad in con-tinuity with the germinal epithelium.

189

The Testis and Epididymis. Fig. 2

In the male the cords are cut off from the germinal epithelium by a layer of mesenchyme, later thickened to form the tunica albuginea. The cords hollow out to form the seminiferous tubules and rete testis. The mesonephric glomeruli degenerate but some of the tubules placed near the rete become connected with it to form the efferent ducts of the testis. These ductules become tortuous and form the main bulk of the head of the epididymis. The mesonephric duct itself, also tortuous near the testis, forms the body and tail of the epididymis. The rest of the duct, leading from the epididymis to the urogenital sinus, forms the vas deferens.

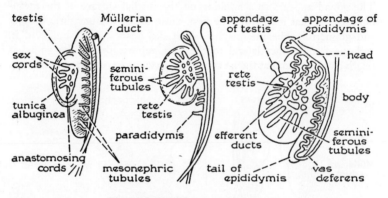

Fig. 18.2. Epididymis and rete testis.

Remnants of the Mesonephric Tubules. Fig. 2

Only a few of the more suitably placed mesonephric tubules are preserved as efferent ductules. Most of the rest, placed too far cranially or caudally to reach the testis, disappear. A few of the cranial tubules may persist as the appendage of the epididymis in the adult, while a caudal group forms the paradidymis. The cranial end of the female duct becomes sealed off in the male, but may persist in the adult as the appendage of the testis.

The Ovary. Fig. 3

In the ovary the sex cords break up and the primordial germ cells become scattered through the mesenchyme. The germ cells develop before birth into young ova, each surrounded by a group of follicular cells to form a primary follicle. In adult life the follicles ripen one by one. There is no strong tunica albuginea and it is possible that new follicles can be formed throughout life by ingrowth from the germinal epithelium, but this is uncertain.

Remnants of Male Ducts in the Female. Fig. 3

The open end of the female duct is preserved as the abdominal opening of the uterine tube, and round the opening the fimbriae are developed as outgrowths. The mesonephric tubules and duct degenerate. But a

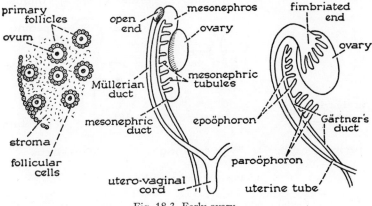

Fig. 18.3. Early ovary.

few tubules opposite the cranial end of the ovary may persist as the epoöphoron and a few of the more caudal ducts as the paroöphoron. The mesonephric duct may persist as Gärtner's duct. Any of these remnants may become cystic in later life and require surgical removal.

The Genital Swellings. Fig. 4

In both sexes two pairs of swellings develop beside the opening of the urogenital sinus. The inner genital swellings are long and narrow and extend onto the phallus beside the urogenital sinus. The lateral genital swellings are broad and short. In the female these swellings persist as the labia minora and majora, but in the male they undergo further development.

The Penile Urethra. Figs. 4 & 5

In the male the inner genital swellings form the urethral folds which extend forwards on the caudal surface of the penis. These unite progressively from the base of the penis towards its tip, cutting off the

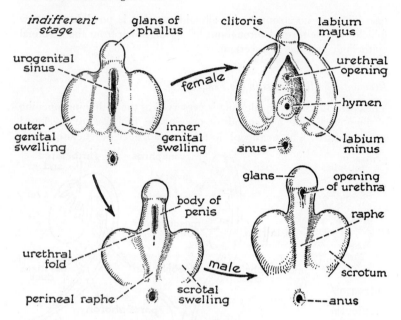

Fig. 18.4. External genital organs.

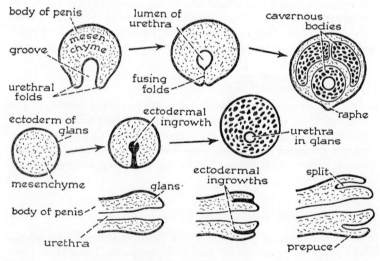

Fig. 18.5. The penis.

penile urethra and carrying its opening towards the glans. Along their line of union they form the perineal raphe, which can be distinguished on the surface in the adult. The glans itself is tunnelled by a special mechanism. A sheet of ectodermal cells grows into its interior from the surface. The attachment of the sheet is lost but its thickened free margin hollows out so as to develop a lumen which carries the urethra forwards to the tip of the glans. Failure of the urethal folds to unite, or of the glans to canalize, gives a penis with the urethra opening on its under surface, the condition of hypospadias.

Prepuce. Fig. 5

A further ring-shaped ectodermal ingrowth forms near the tip of the glans. After birth this splits so as to separate the prepuce from the surface of the glans. The cavernous bodies of the penis and clitoris are developed from the mesenchyme.

Seminal Vesicles and Prostate. Fig. 6

A diverticulum from the mesonephric duct, near its termination in the urethra, gives the seminal vesicle and marks off the vas from the ejaculatory duct. Later numerous outgrowths from the urethra build the prostate, which comes to enclose the ejaculatory ducts. Also enclosed in the prostate is the termination of the Müllerian cord which forms the so-called prostatic utricle. It is really a vaginicle (vagina masculina), as is clear from its development and from its reaction to female hormones injected into male animals. In the female there may be 'paraurethral ducts of Skene' representing the prostate.

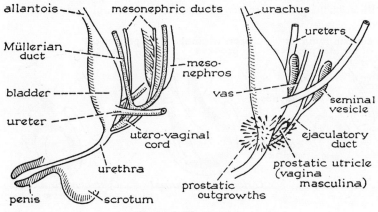

Fig. 18.6. The prostate.

Vaginal Process and Gubernaculum. Fig. 7

In the male the outer genital swellings become the scrotal swellings into which vaginal processes of the peritoneum grow. A cord of condensed mesenchyme, the gubernaculum, forms between each testis and the corresponding swelling, raising a ridge on the posterior abdominal wall. The testis at this stage lies beneath the diaphragm in the upper part of the abdomen.

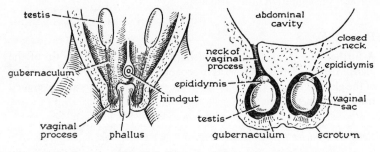

Fig. 18.7. Descent of the testis.

Descent of the Testis. Figs. 7 & 8

The gubernaculum shortens while the rest of body grows, so that the testis is drawn or guided along the dorsal wall of the vaginal process. Before birth the neck of the process closes off, so as to leave the testis behind an isolated pocket of peritoneum, the vaginal sac of the adult. If the neck of the process fails to close off the intestine may enter the scrotum through the open passage, giving congenital inguinal hernia. Local failures of closure may lead to the formation of cysts along the

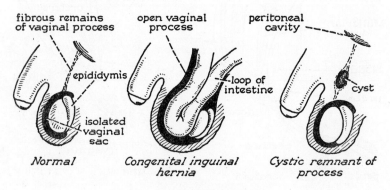

Fig. 18.8. Congenital hernia.

course of the process, and these may enlarge to give hydrocele of the cord. Testes which have failed to descend are usually sterile and secretion of sex hormone may also be insufficient.

Gubernaculum in the Female. *Fig. 9*

In the female the gubernaculum forms and the ovary descends, but not so far as does the testis in the male. The gubernaculum is caught up in the condensed mesenchyme that forms the wall of the uterus, so as to become divided into two parts: the ligament of the ovary passing from the ovary to the uterus and the round ligament of the uterus which ends in the outer genital swelling or labium majus. The ligament of the ovary shortens as the ovary takes up its adult position in the pelvis.

Broad Ligament. *Fig. 9*

As each uterine tube passes towards the mid-plane to join the Müllerian cord it raises a peritoneal fold, the rudiment of the broad ligament. Later the ovary comes to be attached to the dorsal surface of this ligament, and the infundibulo-pelvic fold, raised by the ovarian artery, forms its most lateral part.

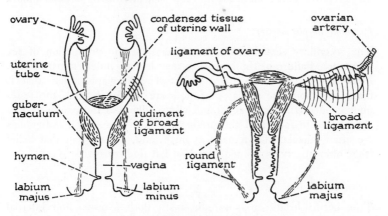

Fig. 18.9. Gubernaculum in the female.

Breast. *Fig. 10*

The breast appears as a local thickening of ectoderm which at first projects from the surface of the body. Later the thickening sinks below the surface, and from it a number of epithelial cords grow into the underlying mesenchyme. This is the state at birth.

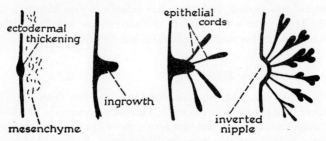

Fig. 18.10. The breast.

Supernumerary Breasts

There are normally only two breasts, as is usual in mammals which bear their young singly. In mammals bearing numerous young the breasts are formed along actual milk lines (mammary ridges) which appear as band-like thickenings of the ectoderm stretching from the axilla to the inguinal region but not in human embryos at any stage. Supernumerary breasts are common in women, especially in the axilla, and may secrete milk. Sometimes, due to the mother's hormones passing the placental barrier, the breast of a new-born infant of either sex may secrete a little milk in the first few days after birth ('witch's milk').

Pseudohermaphroditism

The secondary sexual characters depend on the production of sex hormones. Failure of the hormones may lead to indeterminate states. Most intersexual humans are male pseudohermaphrodites, that is males in whom the sexual characters are poorly defined, the testes undescended, the penis small and clitoris-like and the penile urethra absent due to failure of the urethral folds to fuse. Such individuals are often mistaken for females at birth, and, if happy in that belief (they often marry as women), are best left undisturbed.

Hermaphroditism

True hermaphroditism is rare in man. Both external and internal genitalia show mixed features of both sexes. For instance, there may be a testis on one side and an ovary on the other. Or, undescended testes may be associated with a normal vagina and female external genitalia, but the uterus is absent. Such testes seem to secrete a feminizing hormone and their removal abolishes libido and causes emotional upset. All hermaphrodites are sterile. The aberration in development is frequently associated with abnormality in number and quality of the sex chromosomes.

19

The Skeleton

Mesenchymal Skeleton and Cartilaginous Centres. Fig. 1

Each limb bud differentiates into an upper segment, lower segment, hand or footplate and digits. The mesenchyme condenses to form a skeleton in which the various regions, scapular, humeral, radial, ulnar etc., can be recognized. All these are continuous, for there are no joints in the mesenchymal skeleton. A cartilaginous centre appears for each piece of the skeleton in the interior of the condensed mesenchyme. Cartilage matrix is formed between the cells while the cells themselves round off to become cartilage cells.

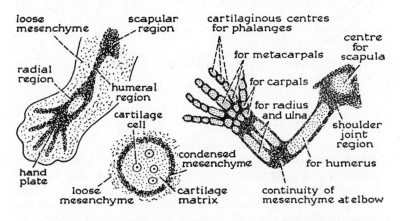

Fig. 19.1. Mesenchymal skeleton and cartilaginous centres.

Interzones. Fig. 2

The centres spread and acquire their characteristic tuberosities, processes etc. to form the cartilage models of the scapula, humerus and other bones. The condensed mesenchyme surrounding each piece

197

forms a perichondrium, and that between the pieces a series of interzones. These interzones are the first rudiments of the joints, and are continuous with the perichondrium.

Three-layered Interzone. *Fig. 2*

In a simple synovial joint the interzone becomes three-layered, with a dense layer covering the cartilage on either side of the joint and a loose layer in between. At the same time the joint capsule appears as a condensation of the general mesenchyme surrounding the joint region. The capsule runs from the perichondrium of one cartilage to that of the other, enclosing the joint region and cutting off intracapsular parts of the perichondrium and mesenchyme from the extracapsular parts.

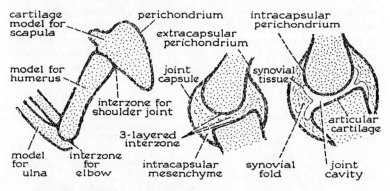

Fig. 19.2. Cartilage models and synovial joints.

Joint Cavity. Fig. 2

Fluid spaces form in the loose middle layer of the interzone and these run together to make the joint cavity. The cavity spreads, following the curvature of the cartilages, cutting out the synovial folds from the intracapsular mesenchyme. Intracapsular ligaments, articular discs, menisci and labra arise as condensations of this mesenchyme, which also forms the synovial tissue of the joint. The dense layers of the interzones become the articular cartilages.

Primary Centre of Ossification. Fig. 3

Near the centre of the shaft of the cartilage model the perichondrium becomes changed into periosteum. The periosteum has two layers, an outer dense fibrous layer and an inner looser osteogenic layer. The osteogenic layer contains osteoblasts and blood vessels. The osteoblasts lay down bone matrix and fibres on the surface of the cartilage model, and these take up calcium salts from the blood to become bone tissue. Thus the primary centre of a long bone takes the form of a cylinder of periosteal bone, open at both ends, surrounding the shaft of the still unbroken cartilage. The edges of the cylinder spread progressively over the cartilage which is later eroded to provide a narrow cavity.

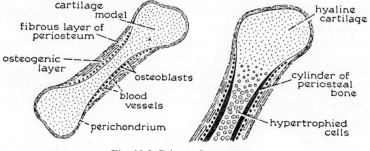

Fig. 19.3. Primary bone centre.

Phocomelia and Amelia. Fig. 4

Reduction and fusion of the major long bones of the limbs, phocomelia (seal limbs), or the entire absence of limbs, amelia, were, and again are, very rare anomalies. But in 1960 and '61 some 7,000 cases of varying

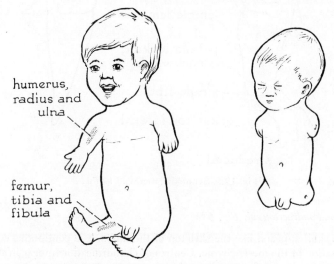

Fig. 19.4. Phocomelia and amelia.

degrees of severity followed the use of thalidomide in treating the vomiting of pregnancy. This is the only known major catastrophe of its kind, but it and the occasional appearance of cleft palate following cortisone therapy have led to much greater care in drug administration to pregnant women.

Sympodia. Fig. 5

Deficiency of tissue at the caudal end of the early embryo gives sympodia. The legs are fused and so rotated that the knees look backwards and the great toes outwards, or they may be reduced to a tapering remnant. The gut ends blindly and the urogenital organs are malformed or absent, apart from the gonads which are more often male than female. Both cyclopia (p. 232) and sympodia appear to have passed into legend, as one-eyed giants and mermaids.

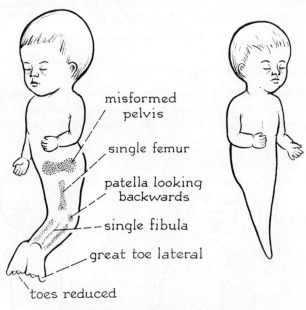

Fig. 19.5. Sympodia, mermaid foetus.

The Chondrocranium. Fig. 6

The skull appears, like the skeleton of the limbs, as a continuous condensation of the mesenchyme. Centres of chondrification appear in this mesenchyme and fuse to form the chondrocranium. The basis cranii lies between the brain and pharynx, in line with the bodies of the cervical vertebrae. Attached to the basis are, in order, the nasal capsules, the lesser wings of the sphenoid, the otic capsules and the occipital arches. The lesser wing is perforated by the optic foramen, while the jugular foramen lies between the otic capsule and occipital arch. The ethmoid bone ossifies in the nasal capsule, the basi-sphenoid and basi-occipital in the basis, and the lesser wing, parts of greater wing of the sphenoid and the jugular process of the occipital in the various processes.

Membrane Bones. Fig. 6

The vault of the skull and the facial bones are ossified in membrane. Paired frontals, parietals, great wings of the sphenoid, squamous parts of the temporal, and a supra-occipital complete the vault, while nasals, lacrimals, zygomatics, maxillae and mandibles build the face. A tympanic ring supports the ear drum. These bones are the remains of an extensive dermal armour which covered most of the body in early vertebrate evolution. The clavicle, the first bone in the body to ossify, is the only other survivor.

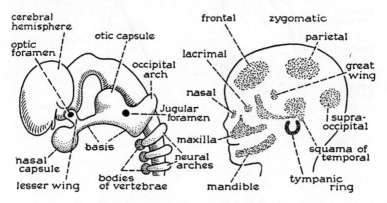

Fig. 19.6. Chondrocranium and membrane bones.

Growth in the Skull. Fig. 7

In foetal life the bones of the vault of the skull are thin and solid. As the brain expands the bone moves out of its way by new bone formation on the outer surface and bone destruction within, as a wall may be moved by building bricks onto one surface and pulling them off the other. (This kind of growth persists after birth in the young pig, and was at one time believed to do so in man. But in fact the growth mechanism changes in childhood, becoming located chiefly at the sutures.)

Skull at Birth. Fig. 7

At birth the face is small compared to the vault, and though the rudiments of the accessory air sinuses are present they are very small. The frontals are paired, separated by the metopic suture which usually closes in childhood, but may persist in the adult. The frontal and parietal bosses, relics of an earlier stage when the head was smaller and the bones more sharply curved, are prominent. There is a large anterior fontanelle, filled by membrane, between the frontals and parietals, and there are other smaller fontanelles. The bones of the vault are thin and one-layered, for there is, as yet, no diploë, and the sutures are loose so that the bones can override each other in childbirth. The jaws are small and filled with developing teeth, not yet erupted, and the ramus and angle of the mandible are poorly developed.

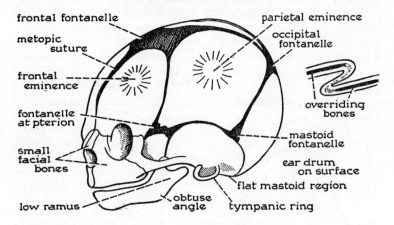

Fig. 19.7. The skull at birth.

Temporal Bone. Fig. 7

Four separate elements of the temporal have been mentioned, the petrous part ossified in the cartilage of the otic capsule, the styloid process belonging to the skeleton of the second branchial arch, the squama ossified in the mesenchyme of the skull vault, and the tympanic ring supporting the ear drum. In most animals these parts remain separate throughout life, but in man they fuse. The ring, the ossicles and the labyrinth are formed full size from the first for the acoustico-vestibular apparatus is delicate and might be disturbed by growth. At birth the drum and tympanic ring lie on the surface of the skull, but later a bony part of the external auditory meatus is developed by outgrowth from the ring. The mastoid process, with its accessory air cells, also grows from the petrous part after birth.

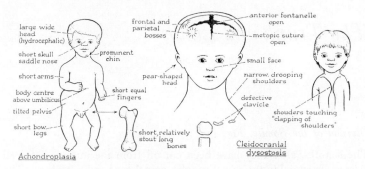

Fig. 19.8. Achondroplasia and Cleidocranial dysostosis

Achondroplasia and Cleidocranial Dysostosis. Fig. 8

These are both genetically determined malformations compatible with long life. In achondroplasia the cartilage bones, apart from the vertebrae, are short and relatively stout. A short skull base gives a saddle nose and deformation of the brain often gives difficulty in the evacuation of cerebrospinal fluid resulting in hydrocephaly. The arms and legs are short so that the umbilicus is well below the body centre. Intelligence is normal. In cleidocranial dysostosis the membrane bones of the vault of the skull are poorly formed and separated by wide sutures and fontanelles and the face is small giving a pear-shaped head. The clavicles, also membrane bones, are deficient or absent so that the shoulders can be brought together in front of the chest.

The Trunk and Limbs

Division of the Somite. Fig. 1

The somites, after they have been cut off from the intermediate cell masses, lie lateral to the spinal ganglia, spinal cord and notochord. The lateral part of each somite expands to form the myotome, which contains the myocoele, a cavity which was originally an extension of the coelom. The medial part of the somite, the sclerotome, grows towards the notochord, between the spinal cord and aorta.

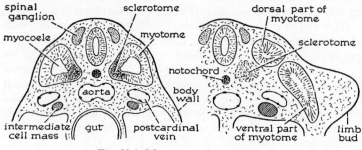

Fig. 20.1. Myotome and sclerotome.

Myotome. Fig. 1

The myotome is cut off completely from the sclerotome and the myocoele obliterated. It becomes divided into a dorsal and ventral part, separated by a septum of ingrowing mesenchyme. The myotomal tissue does not enter the limb buds, but the ventral part spreads into the body wall.

The Myotomal Musculature. Fig. 2

The myotomes become rearranged and fused with each other to give the muscles of the trunk. The dorsal parts form the back muscles, that is, those supplied by the posterior primary divisions of the spinal nerves, including the suboccipital muscles. The ventral parts give the prevertebral and infrahyoid muscles, the intercostals and the muscles of the abdominal wall, all supplied by anterior primary divisions. The rectus abdominis is formed at first in the mid-axillary line, and only later migrates to the front of the abdomen, strengthening the previously membranous abdominal wall. The inscriptions of the rectus may be the remnants of the mesenchymal septa between the myotomes.

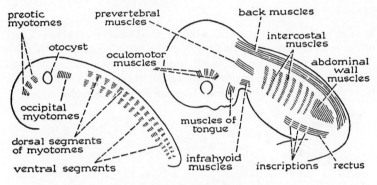

Fig. 20.2. Myotomal musculature.

Eventration of Viscera and Sternal Fissure. Fig. 3

Before the myotomal musculature has migrated ventrally the thoracic and abdominal walls are very thin (figs. 9.10 and 9.11). If they break down the viscera protrude, giving ectopia cordis or eventration of the abdominal viscera. The sternum appears as two sternal bars which later fuse with the ends of the costal cartilages and with each other. Failure of fusion gives sternal fissure or perforate sternum, common and usually harmless anomalies.

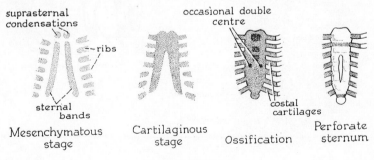

Fig. 20.3. Sternum.

Head Myotomes. Fig. 2

The head myotomes, at least partly derived from prochordal plate tissue, make the oculomotor muscles. Three preotic masses form those supplied by the oculomotor, trochlear and abducent nerves. Occipital myotomes from behind the otocyst migrate into the tongue to form the musculature supplied by the hypoglossal nerve.

Muscles of Limbs. Fig. 4

If a few myotomes are removed from an amphibian larva there is a gap in the musculature of the body wall in the resulting adult. This confirms the derivation of the trunk musculature from the myotomes. The limb muscles are, on the contrary, intact, for they are developed from the mesenchyme of the limb bud, not from the somites.

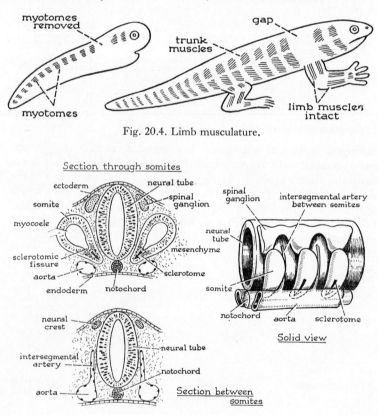

Fig. 20.4. Limb musculature.

Fig. 20.5. Formation of sclerotomes.

The Sclerotome. Figs. 1 & 5

As the sclerotome separates from the myotome it carries with it an extension of the myocoele, the sclerotomic fissure. At this stage the neural tube, notochord and endoderm are still in close contact so that the sclerotomal tissue cannot spread between them. The aorta is still paired, each vessel giving a series of intersegmental arteries which run over the neural tube between the somites.

The Mesenchymatous Vertebrae. Fig. 6

The neural tube and endoderm move away from the notochord allowing the sclerotomes of either side to meet around it. The sclerotomes become joined along the length of the notochord by a perichordal cylinder whose cells are circularly arranged. Thus the vertebral column is now a

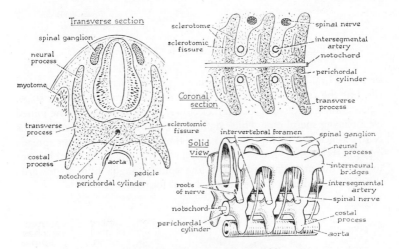

Fig. 20.6. The mesenchymal column.

continuous structure from end to end, though the original segments are marked by alternate denser and looser regions of the mesenchymal blastema. From the lateral margins of the somite three processes grow, a neural process dorsally enclosing the neural tube and spinal ganglia, a short transverse process laterally and a costal process ventrally on either side of the aorta. The neural processes are joined by interneural bridges, so cutting off the intervertebral foramina which transmit the spinal nerves and intersegmental arteries.

Cartilaginous Centres. Fig. 7

A pair of cartilaginous centres, one on each side of the notochord, soon join to surround it and form the body of a vertebra. The body expands rapidly, pushing the intersegmental artery close to the spinal nerve. The dense unchondrified tissue remaining between the bodies becomes the

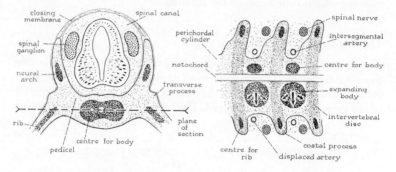

Fig. 20.7. Cartilaginous centres for vertebrae and ribs.

intervertebral disc. A centre in the costal process forms the cartilaginous rib, but the transverse process remains mesenchymatous at this stage. The neural processes spread rather slowly but their tips become joined by a closing membrane which defines the spinal canal. A cartilaginous centre appears in each process to form the neural arch and later spreads into the transverse process and pedicle, also into the interneural bridges to form the articular processes. The spinal ganglia migrate towards their adult positions in the intervertebral foramina.

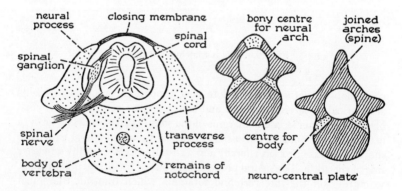

Fig. 20.8. Cartilage model and ossification of vertebra.

Ossification. Fig. 8

The cartilaginous centres join to complete the cartilage model, still open posteriorly except for the closing membrane. Three bony centres appear, one for the body and one for each neural arch. Those for the neural arch are triradiate, with processes extending into the pedicle, the transverse process and the lamina, the last meeting posteriorly to build the spine. After birth the neural arch centres join, leaving only the neuro-central cartilages as growth plates. In the body the notochord is destroyed but remnants persist in the nucleus pulposus of the inter-vertebral disk.

Cervical Rib. Fig. 9

In the cervical vertebrae the costal element, instead of forming a rib, makes the anterior bar of the vertebrarteral foramen and anterior tubercle of the transverse process, the posterior bar and tubercle representing the transverse process of a thoracic vertebra. In the 7th and sometimes the 6th cervical the costal element has its own ossification centre. The 7th element may develop into a separate cervical rib, usually on one side only. Such an extra rib may give no trouble, but the lower roots of the brachial plexus must cross it and, in the case of T1, hook sharply over it or over its fibrous continuation to the clavicle, and pressure symptoms may necessitate its surgical removal.

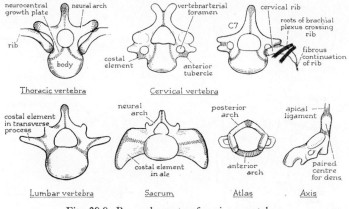

Fig. 20.9. Bony elements of various vertebrae.

Particular Vertebrae. Fig. 9

Paired centres corresponding to neural arches ossify the lateral masses and posterior arch of the atlas, but leave the anterior arch fibrocartilaginous at birth, later to be ossified from a separate centre. The axis has paired centres for the dens in addition to the centres for the neural arches and body. The dens is commonly said to be the body of the atlas attached to the axis, but this is not strictly true, for at the time the atlanto-axial joint was evolved there was no axis to have a body. It is, however, true that the material, which, if it had been situated further from the head, would have formed the body of a vertebra has indeed become a part of the axis. The apical ligament of the dens is the only part of the notochord to persist into adult life apart from the remnants in the intervertebral discs. In the lumbar region the costal elements ossify separately as the 'transverse processes' and in the sacrum, where five originally separate vertebrae coalesce, the costal elements build the greater part of the alae.

Spina Bifida and Myelocoele. Fig. 10

Mid-line defects of the back due to failure of full fusion of the neural folds or neural arches of the vertebrae are relatively common. Simple bone defects, in which the neural arches have failed to meet so that the

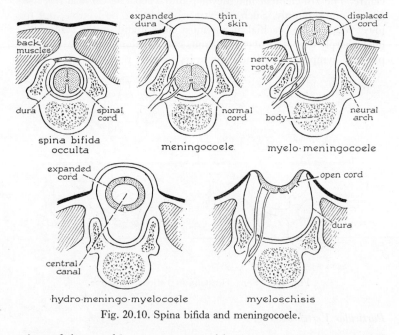

Fig. 20.10. Spina bifida and meningocoele.

spines of the vertebrae are represented by two separate processes, and the gap is filled by membrane and covered by skin (spina bifida occulta), give little trouble. But the meninges alone (meningocoele) or the spinal cord in addition (myelo-meningocoele) may bulge through the gap, or the central canal of the cord may be expanded, hydro-myelo-meningocoele, or in the most severe condition, myeloschisis, it may open on the body surface. These defects of the cord and meninges are necessarily accompanied by spina bifida. The seriousness of such defects depends on the amount of damage to the cord.

Muscles, Tendons and Joints. Fig. 11

The limb muscles appear soon after the mesenchymal skeleton as further condensations, the premuscle masses, one for each main flexor or extensor group. These split into separate condensations for the individual muscles, the tendons being at this time indistinguishable from the muscle bellies. The muscles spread to become attached to the cartilaginous epiphyses or to the periosteum over the shafts of the bones, and later differentiate into muscle bellies and tendons.

Foetal Movements

At the end of the 2nd month, about the time the middle layers of the interzones break down and the motor nerves reach the muscle fibres, the long muscles of the neck and trunk begin to contract spontaneously and later arm and then leg movements are added. By 3 months some postural reflexes are established and the foetus reacts to skin stimulation. The movements may be felt by the mother (quickening) or by the physician through the abdominal wall during the 5th month of pregnancy, and these exercises are probably essential to proper muscle and joint development.

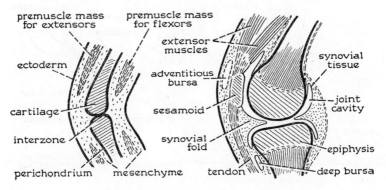

Fig. 20.11. Muscles, tendons and joints.

Deep Bursae. Fig. 11

The deep bursae, such as the iliopsoas, biceps and semimembranosus bursae, appear as splits in the mesenchyme at about the same time as the cavities of the joints near which they lie. Their walls undergo synovial transformation so as to become indistinguishable from the joint linings. They are, from the beginning, separate from the joints. (But in later life the tissue separating them may wear through. Thus the hip joint may, in the adult, come to communicate through its capsule with the iliopsoas bursa, but is never found doing so in the newborn.)

Adventitious Bursae. Fig. 11

In foetal life there is no sign of the subcutaneous bursae which, when inflamed, give 'parson's knee', 'writer's elbow', 'miner's hip' and similar complaints. They are formed after birth in response to rubbing of the skin over the underlying tissues at the olecranon, knee, hip and other bony prominences.

Arteries of the Arm. Fig. 12

The axis artery persists as the subclavian-axillary-brachial trunk, continued in the early embryo to the capillary plexus of the palm. Later two new channels develop as branches from the axis artery, the median, which accompanies the median nerve, and the posterior interosseous. The median now becomes the main artery supplying the palm, while the axis artery becomes the anterior interosseous. Later still the radial and ulnar arteries develop and take over the supply of the palm. The

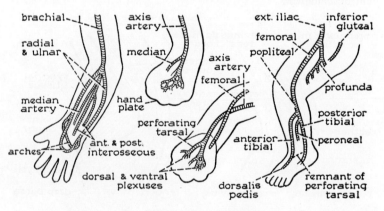

Fig. 20.12. Arteries of limbs.

palmar arches are formed from the original palmar plexus. Sometimes development is arrested, and the median or anterior interosseous remains the most important artery in the forearm of the adult.

Arteries of the Leg. Fig. 12

In the leg the axis artery lies with the sciatic nerve. Reaching the foot it supplies the sole, and, through a small branch, the perforating tarsal, which passes through the tunnel between the talus and calcaneum, the dorsum. An axis artery in the posterior compartment of the thigh would be liable to accidental compression in the adult, so it is replaced by the femoral. A new channel, the profunda, supplies the posterior compartment by its perforating branches. The femoral is continued to the foot as the posterior tibial. The anterior tibial is a new artery which takes over the supply of the dorsum of the foot when the perforating tarsal disappears. The remains of the axis artery make the inferior gluteal and its sciatic branch proximally and the peroneal distally.

Veins of the Limbs

The capillary plexuses of the early limb buds drain into veins along the cranial (preaxial, with the thumb or great toe) and caudal (postaxial) borders. These veins persist as the cephalic, basilic, great saphenous and small saphenous veins, all at first tributaries of the cardinal veins but later finding new exits.

Limb Anomalies

The causes of club foot, the commonest serious anomaly of the limbs, are unknown. In mice it is caused by an excess of fluid expelled from the midbrain, a 'mesencephalic bleb', travelling to the extremity of the limb and there disturbing development, but not in man. Congenital dislocation of the hip is due to a failure of ossification in the rim of the acetabulum, the effect of a recessive gene. The joint does not always dislocate even when the bone is deficient and in later growth the deficiency is made good. In Africa congenital dislocation is very rare, perhaps because the gene is rare, perhaps because babies are carried, using a special carrier cloth, on their mothers' backs with their legs well apart, the position in which infants with dislocation are treated in Europe.

The Nervous System

Autonomic System. Figs. 1 & 2

The autonomic system is built of cells migrating from the neural crest to form the sympathetic and parasympathetic ganglia. A group of neurons of the basal lamina of the cord near the limiting sulcus makes the viscero-motor column. The axons of these neurons grow out by the ventral root to enter the spinal nerve, and leave it as a white ramus to an autonomic ganglion. The axons of the ganglionic cells re-enter the nerve as a grey ramus. The coeliac and other peripheral sympathetic ganglia contain neurons which migrate further. The ciliary ganglion, the most cranial of the parasympathetic series, is made by an anterior extension of the main neural crest derived from the optic vesicle, the other parasympathetic ganglia of the head are derived from neurons migrating from the early facial and glossopharyngeal ganglia. The ganglion cells of myenteric plexus of the intestine, both small and large, come from

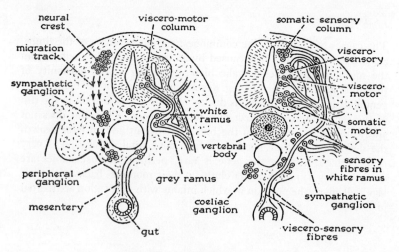

Fig. 21.1. Sympathetic nerves.

the vagus ganglion and migrate down the length of the gut. If they fail to reach the lower part of the colon and rectum that part cannot relax so that the intestine above it becomes grossly distended giving Hirschsprung's disease.

Viscero-sensory Neurons and Nerve Components. Fig. 1

Some of the cells of the spinal ganglia develop processes which spread along the sympathetic nerves to reach the viscera. These behave as do the nerves of common sensation, but have their own termination in the grey matter of the alar lamina, the viscero-sensory column. Thus the spinal nerves have four components, two sensory ending in the alar lamina, the somatic sensory and viscero-sensory, and two motor, viscero-motor and somatic motor, in that order.

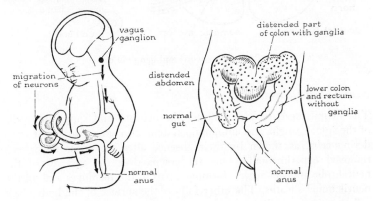

Fig. 21.2. Migration of neurons of intestinal plexuses and Hirschsprung's disease.

Unipolar Neurons. Fig. 3

The cells of the early spinal ganglia are bipolar, with a process growing from each end, one to bring sensory impulses to the cell and the other to pass them on to the cord. Later the paired processes spring from a common stalk, so that the cells become unipolar. The impulses are now not delayed by having to pass through the cell body.

Shape of the Cord. Fig. 3

The early cord has a large central canal, and an alar and basal lamina of equal size. As the motor cells develop the basal lamina becomes for a time much the larger. Later still the alar lamina also enlarges. The two sides of the cord expand more than the central part. So the canal is reduced and the sides are separated by a dorsal raphe and a ventral sulcus. The viscero-motor column forms the lateral horn in the adult, but the viscero-sensory column is not so demarcated.

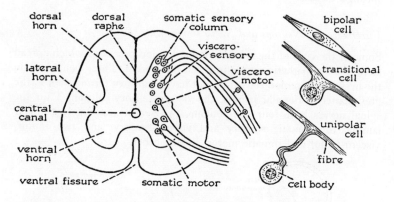

Fig. 21.3. Spinal cord and ganglion cells.

Neural Crest Derivatives

The neural crest, besides forming the neurons and supporting cells of the spinal ganglia and autonomic system, also emits two sets of wandering cells, as shown by their absence after the crests have been removed experimentally. One set creeps over the surfaces of the neural tube and the nerves, forming a part of the meninges and all the neurilemmal sheaths. The other follows the surface of the body, each cell secreting a repellant to induce scattering. Eventually the cells settle

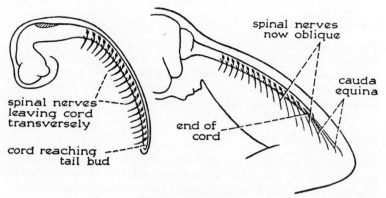

Fig. 21.4. Recession of cord.

in the deeper layers of the epidermis and manufacture pigment which is passed on to the skin and hairs. In albinos and whites the melanocytes are less active than in negroes, though just as numerous. Even in negroes activity increases after birth, so that the new-born baby is relatively light coloured.

Withdrawal of the Cord. Figs. 4 & 5

In the early embryo the cord extends through the whole length of the spinal canal. Later it does not grow as fast as the rest of the embryo so that its caudal end is withdrawn to the upper lumbar region. The roots of the spinal nerves become oblique, the most caudal forming the cauda equina.

Arnold-Chiari Malformation. Fig. 5

Should the cord be unable to withdraw, as when it is fixed by an un-closed neuropore or neurenteric canal, it still contracts. This drags the brain downwards through the foramen magnum, so that the whole of the medulla and part of the cerebellum, all tightly compressed together, are found in the neck, the Arnold-Chiari malformation. The fourth ventricle is greatly elongated but the compression blocks the exits for the cerebrospinal fluid, so there is associated hydrocephalus.

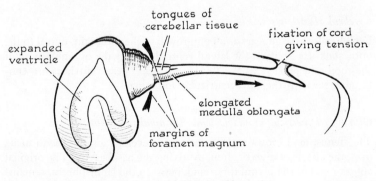

Fig. 21.5. Arnold-Chiari malformation.

Parts of the Brain. Fig. 6

As the cephalic flexure deepens the parts of the brain become more distinct. The hindbrain, with its thin roof, is still the longest part cut off by its isthmus. The midbrain is sharply bent round the flexure, so it has a long roof, but it is distinctly narrower than the fore- or hindbrain.

The forebrain is marked off by two small diverticula, the rudiment of the pineal gland and the rudiments of the paired mamillary bodies.

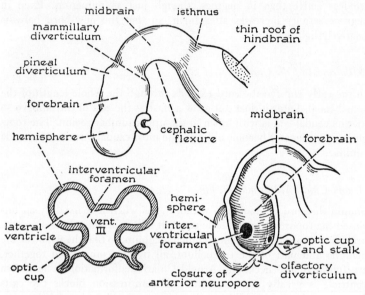

Fig. 21.6. Cerebral hemisphere.

Cerebral Hemispheres. Fig. 6

Each hemisphere appears as an outgrowth of the forebrain, dorsal to the optic stalk. Its cavity is the lateral ventricle, while the original cavity of the forebrain persists as the third ventricle. These ventricles are continuous through the interventricular foramina.

Olfactory Diverticulum and Lamina Terminalis. Fig. 6

The hemisphere grows rapidly and soon develops an outgrowth of its own, the olfactory diverticulum, overlying the nasal cavity. The original anterior end of the neural tube, sunk between the hemispheres, remains thin as the lamina terminals, with one very thin spot marking the place of closure of the anterior neuropore.

Thalamus and Corpus Striatum. Fig. 7

The thalamus develops as a thickening of the wall of the caudal part of the forebrain which bulges into the third ventricle. Later, the hemisphere spreads so as to cover the thalamic region. Here the wall of the

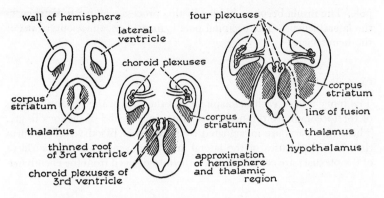

Fig. 21.7. Corpus striatum.

hemisphere thickens to form the corpus striatum. Both the thalamus and the corpus striatum are made of nerve cells of the mantle layer. Ventral to the thalamus a smaller thickening of the forebrain makes the hypothalamus.

Development of the Hemisphere. Fig. 8

The cranial end of the hemisphere becomes the frontal pole. The olfactory diverticulum is originally hollow, but later its walls thicken and obliterate the cavity. Its swollen end, which receives nerve fibres growing into it from the olfactory epithelium, becomes the olfactory bulb, and its stalk the olfactory tract. The part of the hemisphere overlying the corpus striatum becomes the insula, which acts as a relatively fixed region round which the hemisphere grows. The caudal end of the hemisphere turns ventrally and finally cranially to form the temporal

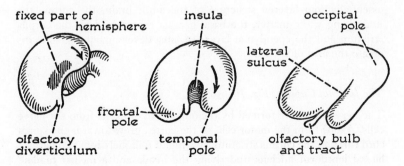

Fig. 21.8. Expansion of hemispheres.

pole. The insula becomes hidden in this process, coming to lie deep in the lateral sulcus. The occipital pole is formed by a new outgrowth of the hemisphere.

Choroid Plexuses. Fig. 7

The inner wall of each hemisphere and the roof of the thalamic region of the forebrain are thinned so as to become devoid of nervous tissue. These thin parts are invaginated by plexuses of blood vessels to form the choroid plexuses of the lateral and third ventricles. The surfaces of the plexuses are covered with epithelium derived from the ependyma.

Cerebral Cortex. Fig. 9

In early development the brain wall is constituted as that of the spinal cord, with three layers, ependymal, mantle and marginal. Later many of the cells of the mantle layer migrate through the marginal layer to take up a superficial position and form the cerebral cortex. This gives the

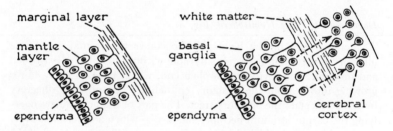

Fig. 21.9. Wall of the hemisphere.

essentially four-layered structure of the adult brain, the cortex, the underlying white matter, the basal ganglia represented by the corpus striatum and the ependymal layer. In later development the sulci increase the area of cortex.

The Internal Capsule. Fig. 10

The nerve tracts are formed by the outgrowth of axons from the nerve cells. Axons from the motor cells of the cortex grow towards the spinal cord through the corpus striatum, splitting it obliquely into two parts, a lateral lentiform nucleus underlying the insula and a medial caudate nucleus next the thalamus. These fibres form a part of the corona radiata, internal capsule and crus cerebri and are continued below as the pyramidal tract. The division of the corpus striatum is not quite complete, for even in the adult there is some continuity between the cranial ends of the lentiform and caudate nuclei.

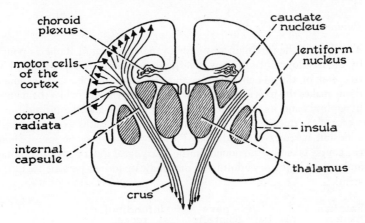

Fig. 21.10. Internal capsule.

Porencephaly. Fig. 11

In the early embryo (figs. 9.9 and 22.1) the neural tube and its derivatives are avascular, being nourished by diffusion from the capillaries of the mesenchyme, but later vessels invade the neural tissue. The tissue is, however, peculiarly sensitive to vascular failure, and blockage of vessels

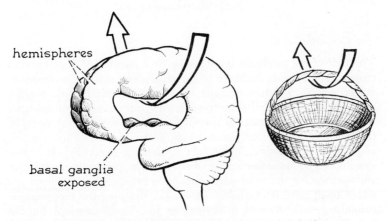

Fig. 21.11. Porencephaly and comparison with a basket.

causes widespread cellular death. In porencephaly, the 'basket brain' of German pathologists, symmetrical necrosis of the hemispheres may leave a wide opening passing right through the brain from side to side, with the basal ganglia in its floor, yet the subject may live.

Pituitary Gland. Fig. 12

The floor of the forebrain responds to the approach of Rathke's pouch by the development of a new diverticulum, the neurohypophysis. The free end of the pouch enlarges and the stem atrophies. The pouch tissue makes the pars distalis, tuberalis and intermedia, while the neurohypophysis and its cavity form the pars nervosa and the infundibular recess of the third ventricle. At birth the cavity of Rathke's pouch is still wide but narrows in childhood. Pituitary tissue may be found along the course taken by the pouch, in the pharyngeal wall or sphenoid bone, and a canal in the dry adult bone may indicate former persistence of the stem.

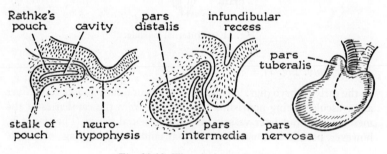

Fig. 21.12. The pituitary gland.

The Commissures. Fig. 13

The roof of the forebrain is thickened by a number of fibre bundles crossing from one hemisphere to the other. The anterior commissure crosses near the lamina terminalis. Fibres of the fornix skirt the margin of the hemisphere, and some cross in the hippocampal commissure. The habenular and posterior commissures cross at the attachments of the pineal stalk. The corpus callosum is developed late (it is a late acquisition in evolution, found only in placental mammals). It begins as a thickening in the forebrain roof above the hippocampal commissure and spreads for the most part caudally as more and more fibres are added to it, until it overhangs the pineal, dragging the hippocampal commissure caudally with it. Sometimes no callosum develops, but the individual, possibly working with only half his brain, is unaffected.

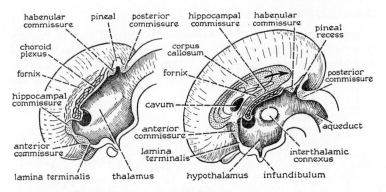

Fig. 21.13. Corpus callosum and third ventricle.

Tela Choroidea. Figs. 13 & 14

As the corpus callosum spreads it cuts the fornix from the rest of the hemisphere. It also delaminates, between itself and the thalamus, a horizontal sheet of vascular mesenchyme, the tela choroidea. This carries the choroid plexuses of the lateral ventricles at its margins, and those of the third ventricle on its ventral surface. There is no nervous tissue in this part of the roof of the third ventricle, so the tela itself, covered only by a layer of ependyma, makes the roof. The septum lucidum may be made from parts of the adjacent medial walls of the hemispheres cut off by the corpus callosum, but this is uncertain.

The Third Ventricle. Figs. 13 & 14

The third ventricle is narrow from side to side, but long and deep, with the thalamus and hypothalamus in its lateral wall. The bulges made by the two thalami usually meet in the median plane and fuse, but no nerve fibres pass through the connexus, so it is not a commissure.

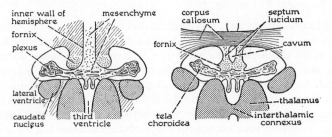

Fig. 21.14. Tela choroidea.

Cranial Nerve Components. Fig. 15

Somatic motor (S.M.) and visceromotor (V.M.) and somatic sensory (S.S.) and viscerosensory (V.S.) nerve fibres and cell columns are found as in the spinal region, but there are three extra columns, acoustico-vestibular (A.V.) and gustatory (G., for taste) in the alar lamina, and

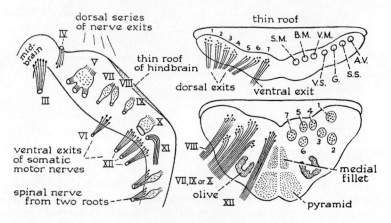

Fig. 21.15. Cranial nerves.

branchiomotor (B.M.) in the basal lamina. Of these components only the somatic motor fibres, which supply the muscles derived from the head somites, emerge as ventral roots. They form the oculomotor, abducent and hypoglossal nerves. The trochlear also belongs here, but it emerges from the dorsal surface of the isthmus, not in line with the other nerves of its group; why, is an unsolved problem of vertebrate evolution.

The Branchial Nerves. Fig. 15

The branchiomotor fibres supply the branchial muscles and the visceromotor form the preganglionic fibres of the cranial parasympathetic system. In the hindbrain region they join with the sensory fibres to form the trigeminal, facial, glossopharyngeal, vagus and accessory nerves. Later fibres growing towards the spinal cord from the cortex build the pyramids, the medial fillet is formed by ascending fibres and the olive appears as a new grey mass between the dorsal and ventral exits. These structures change the shape of the hindbrain but the seven grey columns remain essentially unchanged.

Pontine and Cervical Flexures. Fig. 16

The cephalic flexure develops early as the head is bent ventrally during folding off. For a time the nervous system grows faster than the rest of the body and the brain stem is thrown into a zig-zag, with two new flexures, a pontine convex ventrally and a cervical convex dorsally. The thin roof of the fourth ventricle is folded and shortened in the process.

The Cerebellum, Pons and Medulla Oblongata. Fig. 16

The cerebellum appears as a thickening of the hindbrain roof between the isthmus and the thin part. It becomes dumb-bell shaped, with a depressed central part and expanded lobes, but in later development the more superficial nerve cells migrate deeply, leaving the cortex devoid of cell bodies. Since, however, there is little myelin it remains grey. Cerebellar cortex is developed in the same way as cerebral cortex. The roof of the isthmus becomes the superior medullary velum. The thin hindbrain roof is invaginated by the choroid plexus of the fourth ventricle. The floor of the cephalic end of the hindbrain is crossed by fibres passing transversely to the cerebellum and becomes the pons, while the rest forms the medulla oblongata.

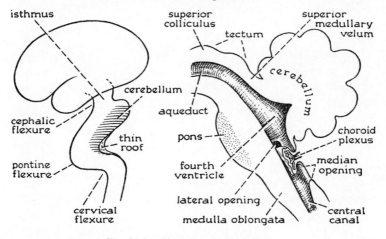

Fig. 21.16. Flexures and cerebellum.

Mid Brain. Fig. 16

The midbrain is at first relatively thin walled, with a roomy interior which continues the cavity of the third ventricle. With the ingrowth of fibres passing to and from the forebrain and the development of the

I

red nucleus the walls are thickened and the lumen reduced to the aqueduct. The roof forms the tectum on which grow the four colliculi.

Escape of Cerebrospinal Fluid. Fig. 16

The thin roof of the fourth ventricle is bulged by the pressure of the fluid secreted by the choroid plexuses, and finally breaks down in three places, the large median foramen of Majendie and the smaller paired foramina of Luschka. Intrauterine inflammations due to infection by organisms that have passed the placental barrier may lead to scar tissue closing these foramina or the aqueduct, with resulting 'congenital hydrocephalus'.

The Meninges, Eye, Ear and Skin.

The Vascular Coat. Fig. 1, A & B

Cells migrate from the neural crest along the surface of the neural tube and mingle with others moving from the somite. These cells together form the first covering of the tube. The development of capillaries over the surface of the tube gives, with this covering, a vascular coat, not yet a true pia (fig. 9.3, 'vascular plexus').

The Spinal Canal. Fig. 1, B & C

The spinal cord is surrounded on all sides by mesenchyme. As the neural arches of the vertebrae develop and become connected dorsal to the cord by the closing membrane to complete the spinal canal this mesenchyme is isolated. In it lie the cord with the roots of the spinal nerves, and an early condensation of the mesenchyme indicates the denticulate ligament. The dura is also formed from this mesenchyme

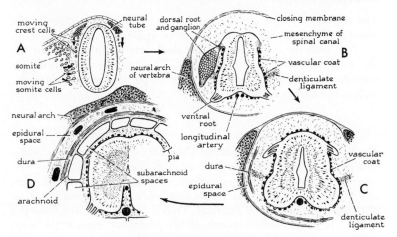

Fig. 22.1. The meninges.

close to the developing vertebrae so that the epidural space is very narrow. The early dura is patchy but the fragments join later to a continuous tube.

The Subarachnoid Space. Fig. 1, D

Cerebrospinal fluid seeping down the cord from the brain splits the loose tissue between the dura and vascular coat to form a network of subarachnoid spaces. The remains of the tissue become added to the inner surface of the dura as the arachnoid and to the outer surface of the vascular coat, where, with the coat, it forms the pia. Later the spaces run together to form a single space with the loss of most of the trabeculae between them. The epidural space widens, becoming filled with loose connective tissue and a venous plexus, and, in the last months before birth, fat.

The Subdural Space. Fig. 1

In the adult the dura can be removed leaving the arachnoid in place. But it is doubtful if the separation actually takes place in normal development, pre- or post-natal. The split is comparable to the split in the decidua that separates the placenta from its bed, the arachnoid trabeculae playing the part of the trophoblastic columns that anchor the plates of the placenta to each other.

Meninges of the Head.

The cranial meninges are developed in the same way as those of the spinal cord, but no fat accumulates in the epidural space. So the dura can fuse with the periosteum of the skull, leaving only room for the venous sinuses and meningeal vessels. The arachnoid trabeculae are numerous, so the arachnoid is always peeled off the deep surface of the dura as the dura is removed.

Lens of the Eye. Fig. 2

The lens vesicle is at first hollow. The cavity is obliterated by elongation of the cells of the deep wall to form the lens fibres, while the anterior wall makes the lens epithelium. The developing lens is vulnerable to the virus of rubella (German measles), becoming opaque instead of transparent (congenital cataract). Thus a mother suffering from the mild fever in the second month of her pregnancy may, if the virus crosses the placental barrier, bear a blind child (see also the inner ear).

The recognition of rubella as an important teratogenic agent was due to an outbreak in Australia followed by the birth of relatively high numbers of malformed babies some months later.

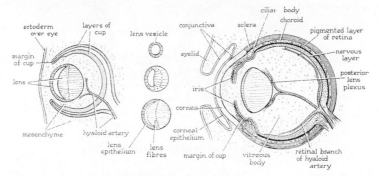

Fig. 22.2. Sclera and cornea.

Retina, Sclera and Cornea. Fig. 2

As the lens develops it moves deep into the optic cup and the margins of the cup grow inwards over its anterior surface. The two walls of the main part of the cup form the inner, thicker, nervous layer of the retina and the outer, thinner, pigmented layer, while the space between the layers is obliterated. The mesenchyme between the lens and the retina becomes the vitreous body, that round the cup the sclera and choroid. That next the ectoderm superficial to the lens condenses to form the cornea, continuous at its margins with the sclera. A deeper part of the mesenchyme lying near the cup margin forms the basis of the iris and ciliary body.

Anterior and Posterior Chambers. Fig. 3

A split in the mesenchyme between the iris and cornea forms the anterior chamber and an independent series of splits between the iris and lens the posterior chamber. The margin of the optic cup, both layers of which become pigmented, forms the posterior surface of the iris. The dark pigment, seen through the translucent body of the iris, gives a blue effect at birth (of white babies), but this is usually modified by other pigments developed in the body of the iris after birth.

The Pupil. Fig. 3

When first formed the anterior and posterior chambers are separated by a complete diaphragm but later this breaks down in the centre to give

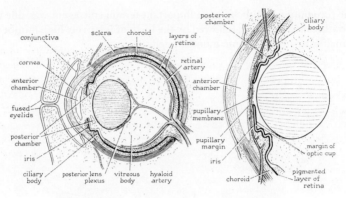

Fig. 22.3. The anterior and posterior chambers.

the pupil. Sometimes remains of the 'pupillary membrane' or of the capillary plexus persist and may, if considerable, interfere with vision.

Hyaloid Artery. Figs. 3 & 4

The hyaloid artery enters the interior of the eye by passing through the choroidal fissure. The fissure closes over, enclosing the artery in the optic stalk. Two sets of branches spring from it, the retinal and the posterior lens plexuses. At birth the eye is already of nearly adult size, and the lens plexus, no longer needed now that growth has slowed, has disappeared. The part of the hyaloid artery that supplied it is reduced to a remnant, the hyaloid canal. The stem forms the central artery of the retina. Occasionally the choroidal fissure fails to close completely, giving coloboma of the adult eye, or the hyaloid artery may persist as an opacity in the vitreous body. The posterior lens plexus may still be visible in premature babies, but soon disappears.

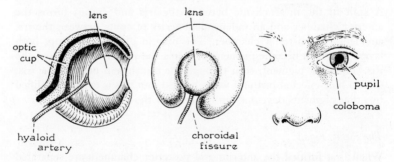

Fig. 22.4. Choroidal fissure and coloboma.

Conjunctiva. Figs. 2 & 3

After the lens vesicle has been cut off, the head ectoderm forms a bulge over the developing eye. A pair of folds forms the eyelids. On the outer surface of each lid the ectoderm develops into skin epithelium and on the inner surface and over the eyeball into conjunctival epithelium. From the conjunctival epithelium develop the rudiments of the lacrimal and tarsal glands. The tissue of the cornea becomes transparent. For a time the lids come into contact and fuse by their epithelial coverings. In man this fusion breaks down before birth, but in many animals, such as cats, the young are born in a less advanced state and the eyes remain closed for some days after birth.

Nasolacrimal Duct. Fig. 5

Where the maxillary process meets the lateral nasal process the surface ectoderm dips in and is eventually cut off as a solid rod of cells. The upper end of the rod grows towards the eye and bifurcates near it to form the canaliculi. A swelling makes the lacrimal sac. An independent outgrowth from the nasal cavity joins the end of the first rudiment. Eventually the canaliculi join the conjunctival sac and the system canalises throughout. Occasionally the rod fails to separate from the overlying ectoderm, and when it opens it leaves a furrow beside the nose, 'oblique facial cleft.'

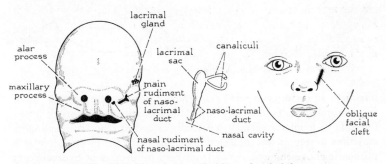

Fig. 22.5. Tear ducts and oblique facial cleft.

Optic Nerve. Fig. 3

The optic stalk is at first hollow and has thin walls. But when nerve fibres derived from the nervous layer of the retina grow into it they thicken the walls and obliterate the lumen. On reaching the forebrain some of the fibres cross in its floor while others remain uncrossed. The two groups build the chiasma and continue as the optic tracts.

Cranial Anomalies. Fig. 6

If the brain is too small (microcephaly) the foetus may survive and grow up, but as an idiot. Too large a brain follows accumulation of cerebro-spinal fluid, usually from blockage of the aqueduct. This results in expansion of the head (hydrocephaly) giving difficulty at birth, and brain damage. Absence of the brain, or a great part of it (anencephaly), due to failure of fusion of the neural folds followed by death of brain tissue and absence of the vault of the skull, is relatively common. Poor development of the nervous mechanism for swallowing may lead to associated hydramnios similar to that caused by a mechanical block in atresia of the oesophagus. The foetus can live for some time after birth, but usually dies quickly from lack of temperature control or from

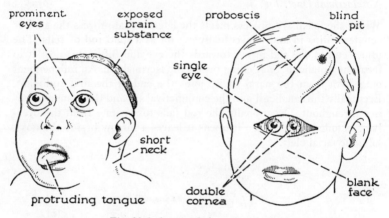

Fig. 22.6. Anencephaly and cyclopia.

infection. In cyclopia an early deficiency of tissue at the anterior end of the forebrain results in a single median orbit with much variety in the form of eye, from none at all to a partially united pair, and corresponding variation in nerves and muscles. The nose is often represented by a proboscis above the eye. Apart from very mild cases of hydrocephaly which can be treated surgically these anomalies are of little clinical interest.

The Inner Ear. Fig. 7

The early otocyst consists of a simple vesicle with a single diverticulum, the rudiment of the endolymphatic duct. The semicircular canals appear as three flanges with thickened borders. Each flange is perforated in the middle to give a tube opening at both ends into the

otocyst, two of the tubes having a common opening. A swelling at one
end of each tube forms the ampulla. A coiled outgrowth from the otocyst gives the scala media of the cochlea. The otocyst itself becomes
subdivided into a saccule and utricle. Condensations of the mesenchyme around form the membranous labyrinth, the ectoderm of the
otocyst giving the epithelial lining. Rubella may cause serious anomalies
of the inner ear, later resulting in deaf-mutism.

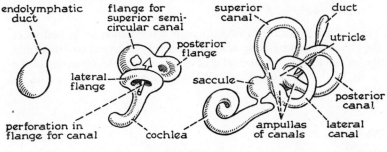

Fig. 22.7. The otocyst.

The Perilymphatic Spaces. Fig. 8

The petrous part of the temporal bone forms around the membranous
labyrinth, giving the bony labyrinth. The loose mesenchyme remaining
between the bony and membranous labyrinths breaks down to give the
perilymphatic spaces, and the scala tympani and vestibuli of the cochlea.

The Middle Ear. Fig. 8

The ossicles are formed early as part of the skeleton of the first and
second branchial arches. They are included by the growth of the
petrous temporal in a space filled with loose mesenchyme. The free end

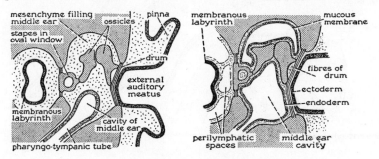

Fig. 22.8. The middle ear.

of the pharyngo-tympanic tube, expanded to form the first rudiment of the cavity of the middle ear, also lies in this loose mesenchyme. Later the cavity extends around the ossicles, at the expense of the mesenchyme, and a further diverticulum forms the mastoid antrum. (After birth accessory air sinuses grow from the antrum into the enlarging mastoid process.)

The Ear Drum. Fig. 8

The drum is formed by a mesenchymal condensation at the bottom of the external auditory meatus. It has an outer ectodermal covering from the lining of the first branchial groove. Later, with the expansion of the cavity of the middle ear, an endodermal epithelium is added on its inner surface.

Nails. Fig. 9

Each nail begins as a specialization of the ectodermal covering of a finger or toe making a nail bed. This pushes beneath the surface as a nail fold. Keratinization of the cells of the fold gives the nail itself, which is forced over the bed so as to reach the tip of the finger at the time of birth.

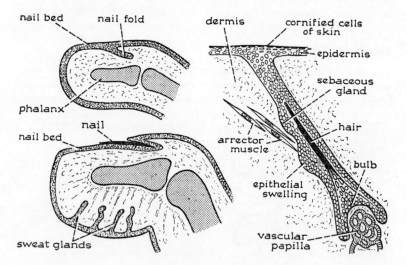

Fig. 22.9. Nails and hairs.

Hairs. Fig. 9

The hair follicle appears as a downgrowth of the deeper layers of the ectoderm into the underlying mesenchyme which is condensing as the dermis. The base of the follicle expands to a bulb; the rest makes the hair sheath. The bulb hollows to enclose a richly vascularized mesenchymal papilla. The basal epithelial cells of the bulb multiply while those nearer the skin surface keratinize as the hair. Two swellings appear on the sheath, one to form the sebaceous gland of the follicle, the other, a temporary structure, probably controlling the development of the arrector muscle from the mesenchyme. The first slender hairs are the lanugo, but most of these are shed before birth.

23
Control of Growth

Embryonic Induction. Fig. 1

A ring of cartilage cut from the suprascapula of an adult frog may be inserted subcutaneously in a tadpole. The tadpole's ectoderm thins and forms no glands within the ring and its mesenchyme condenses, the whole formation suggesting an ear drum. Any cartilage will produce the

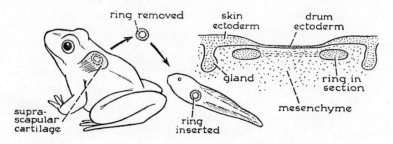

Fig. 23.1. Induction of ear drum.

effect: the suprascapula is merely a convenient source. In the normal embryo the induction of the drum depends on the close proximity of the cartilage to the ectoderm, at the right time, when the ectoderm is responsive, and the reason why all the ectoderm is not of drum type is its distance from the skeleton except in the drum region.

Interaction of Cell Types. Fig. 2

Both mesenchymal cells and cells of the nephrogenic cord can be grown in tissue culture keeping their individual characteristics. The mesenchymal cells are irregular in shape and wander, while the nephrogenic cells are more rounded and spread over surfaces in sheets. But if the two sorts are mixed they build a tissue of a higher order. The epithelial cells make hollow vesicles like the tubules of the early mesonephros, and the mesenchyme comes into orderly arrangement round these vesicles.

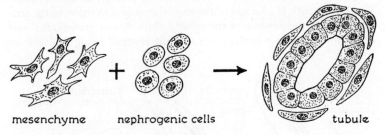

mesenchyme   nephrogenic cells   tubule

Fig. 23.2. Tissue synthesis.

Organization in the Embryo. Fig. 3

In the frog the notochord and endoderm are invaginated at the blastopore, and some of this tissue is responsible for an embryonic hormone (evocator) which causes the neural tube and somites to differentiate. Injection of a few cells from the blastopore of another embryo, even if they are crushed or boiled, may induce a second neural tube and set of somites to differentiate from the tissues of the 'host' embryo. In these and other instances the mechanism of organization is similar. A chemical substance secreted by one tissue acts on another in its immediate vicinity which is prepared to react, to induce a higher degree of organization. Both chemistry and morphology are concerned in the physiology of development.

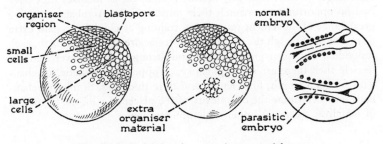

organiser region   blastopore   normal embryo

small cells

large cells   extra organiser material   'parasitic' embryo

Fig 23.3. Induction by organiser material.

Muscle Fibre Length. Fig. 4

In any individual the length of the muscle fibres is so adjusted that the length of a fully contracted fibre is a little less than half its length when fully stretched. Thus the sternomastoid, which has a great range of action, has fibres which run from end to end, while the subscapularis, whose insertion on the lesser tuberosity of the humerus has a restricted range, has correspondingly short fibres in multipennate arrangement. The fibre length is adjusted to the movements by extension of the tendon fibres at the expense of the muscle fibres during growth, any part of the muscle fibre which cannot usefully contract being replaced by

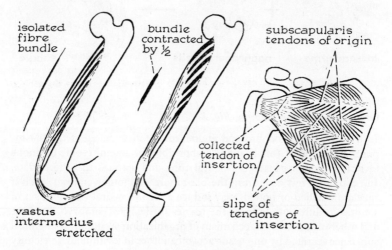

isolated fibre bundle

bundle contracted by ½

subscapularis tendons of origin

collected tendon of insertion

vastus intermedius stretched

slips of tendons of insertion

Fig. 23.4. Muscle fibre arrangement.

tendon. Newborn babies are 'muscle bound' for as foetuses they have had no opportunity to stretch their muscles fully. Acrobats requiring great range of movement must begin training as children to gain extra fibre length. Though the number of fibres is already fixed at birth, each individual fibre can grow thicker and stronger if it is exercised. In muscle then there are self-regulatory, as opposed to genetic, mechanisms for total length, fibre length and fibre thickness. Similar mechanisms may determine the complexity of the placental villi, the formation of new liver tissue and other structural features.

Inhibitory hormones.

Inhibitory hormones secreted by mature cells may prevent overgrowth of formative cells. A secretion of the outer layers of the epidermis, for example stops division in the germinal layer. But as the superficial cells are lost or destroyed the deeper cells divide to make up the loss.

Teratomas, Hamartomas and Misplaced Tissues. Fig. 5

Nests of undifferentiated embryonic tissue may persist in many parts of an otherwise normal body and may later differentiate to form tumours. Teratomas, commonly found in the ovaries and testes but also elsewhere, may show skin, hair, nails, even well formed digits, teeth, fragments of intestine, nervous system or skeleton. This complex was at one time believed to represent an included and disorganised twin, but that suggestion is now discredited. Vascular hamartomas, 'angiomas' and 'birthmarks', lipomas and fibromas, pigmented moles or naevi can be found in almost any individual, all the result of minor errors of development usually quite harmless but occasionally becoming malignant. Fragments of embryonic material that become misplaced probably account for the accessory thyroid, thymus, pancres, spleen and other tissues found scattered in the body, often becoming cystic as their cells secrete. Chemical errors including the malformation or absence of enzymes and hormones and peculiar variations of haemoglobin mostly depend on deficiencies of information in the genetic code.

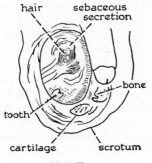

Fig. 23.5. Teratoma.

Times of Development

A student is not expected to remember the times at which the embryo reaches particular stages, but a few points may be indicated (they are taken from Arey's excellent table). At 2 weeks the primitive streak is active, at 3 the somites are distinct. At 4 the branchial arches are developed and at 5 the embryo resembles the 7 mm. pig discussed in chapter 9. At 2 months the face and digits are formed, ossification is beginning and the gut is in the extraembryonic coelom. At 3 the gut has returned and blood formation begun in the bone marrow. At 4 foetal movements have begun, at 5 the lanugo hair is developed and at 6 the foetus can survive premature birth if kept under very favourable conditions.

Anomalies

Anomalous development may occur if genetic (foetal) or environmental (maternal) conditions are adverse. An organ is most vulnerable when its rudiment is being differentiated, and so the 5th and 6th weeks after fertilization are very critical for most organs, the danger receding gradually after the 6th week. Major defects are unlikely to be caused after the 7th week in optic cup derivatives, cardiac valves and limbs (except late amputations), after the 9th week in the hearing apparatus, after the 10th week in branchial arch derivatives, abdominal wall and viscera, after the 11th week in the meninges and spine, and after the 12th week in the pelvic organs.

Onset of Function

The appearance of function in any organ or cell group does not await complete histological differentiation. The most outstanding example of this is the contraction of the endothelial cardiac tube before myoblasts are formed. Voluntary muscle can contract as early as the beginning of the third month, although motor plates form later, during months 5–7. Peristalsis in smooth muscle has been seen early in the third month. Mucus is seen in the intestine in the eighth week, pepsin in the stomach in the sixteenth week (but acid only after birth) and trypsinogen in the pancreas also at about the same time. Endocrine glands begin functioning early. The anterior pituitary is functional at week 8, adrenal medulla at week 16 and thyroid and pancreatic islets between those periods. The interstitial cells of the testis are active early, possibly to combat the influence of leaked maternal hormones, but those of the ovary show no activity in the foetus. The adrenal medulla is certainly active soon after the pituitary, but the picture in the cortex is confused by the growth of foetal cortex which regresses soon after birth. The metanephros is secreting by the 12th week. Cerebrospinal fluid is secreted in the 2nd month and has escaped into the subarachnoid space in the third month.

Maturity and Postmaturity

The mechanism of onset of birth is not known. Normally the foetus weighs about 5 to 8 lbs., subcutaneous fat is well developed so that the skin is no longer wrinkled, the secondary centres for the femur and tibia at the knee and primary centres for three tarsal bones are present, the eyes are open and light sensitive and the lens plexus gone, and the nails have reached the ends of the fingers. But a grossly premature foetus with none of these developments can survive, given special care, for the respiratory apparatus, taste receptors and digestive mechanisms are capable of function at 6 months. Postmaturity is evidenced by over-weight and ossification in the head of the humerus. Compensatory over-growth of the terminal villi of the placenta may impede the circulation of maternal blood in the intervillous space, leading to death of the foetus. Postmaturity may also cause difficulties during parturition, dangerous to both foetus and mother, but pregnancy can be terminated artificially.

Note on Sources

Most of the material we discuss will be found in the standard textbooks, Arey, Putten or Hamilton, Boyd and Mossman, and these may be consulted for photographs of actual human embryos. It is always worth looking up Keibel and Mall (long out of print) and Frazer, as so much has come down from them. But a few topics may be listed in which we have given unorthodox or less known views, and for these we give our sources.

TIMING: Tables are given in the larger textbooks, and a concise account which can be used in revision of the whole subject is found in the second chapter of R.A.Willis, *The Borderland of Embryology and Pathology*, Butterworth, 2nd ed., 1962. Willis also considers embryonic abnormalities in great detail with many references.

HEART: A revolutionary paper by de Vries and Saunders (1962), Contrib. Embryol., Carneg. Instn., 37, reinterprets the early stages. Streeter's (1951) Developmental Horizons, a book of collected reprints, gives a clear account and beautiful drawings of the later stages. We have used the word auricle without reference to the atrium as students find the distinction difficult and it is of little value.

TROPHOBLAST: Studies by Böving (1959) on the rabbit, Ann. N.Y. Acad. Sci. 80, suggest a simple alkalinity as the cause of embedding, rather than the trypsin-like digestive enzyme formerly postulated, and his results probably apply to man.

HAIRLESS SKIN: The surprising derivation from thickened ectoderm of the early limb bud was discovered by Cauna (1963), J. Anat., 97. The feature runs through the vertebrates and suggests an origin of the limbs as sense organs which came to project from the trunk and later became mobile.

THYMUS: See the review by Levey in Scientific American, July 1964, giving results of its experimental removal in young animals.

FACE AND PALATE: We have followed the account of the surgeon Victor Veau as this has been substantiated by Streeter (Developmental Horizons), abandoning the supposed union of free maxillary and frontal processes. There is in fact at no stage a frontonasal process, and Streeter's drawings make it quite clear that the maxillary processes never end freely. Hare lip is discussed by King (1954), J. Anat., 88, and Abdul Aziz (1959), J. Fac. Med. Baghdad, 1.

NASOLACRIMAL DUCT: Our account is taken from Duke-Elder's System of Ophthalmology, Vol. 3. It differs from that usually given. 'Oblique facial fissures' may be gross fractures of the face, usually associated with other anomalies. with the fracture passing through, not along the nosolacrimal duct. Sensory fibres of the maxillary nerve grow across the duct to supply the side of the nose, ignoring any distinction between the maxillary and 'frontonasal' processes.

MENINGES: Our account is based on figures of Weed (1917) and Sensenig (1951), Contrib. Embryol., 5 and 34, and the crucial figure of cavitation from Hamilton, Boyd and Mossman. There is no simple stain for the ground substance of connective tissue. Ranvier's cross was cut in fresh tissue but he could not see its margins under the microscope: Bensley introduced fluid droplets containing Paramoecia, and while he could not see the droplets the animals changed direction at their presumed margins. This technical difficulty makes us reject Sensenig's account of cavitation before the arrival of cerebrospinal fluid. Again the scarcity of mitoses in the perichondrum of growing cartilages implies recruitment of cells from the surrounding mesenchyme to make up for the known loss to the cartilage, and it seems most unlikely that any cells would be moving in the opposite direction to form the dura. So we have preferred an older view, the condensation of the dura in the mesenchyme of the vertebral canal rather than its splitting off from the perichondrium of the developing vertebrae. Other textbook writers say nothing of the development of the meninges.

DUCTUS VENOSUS: See Dickson (1957), J. Anat., Lond., 91, who confirmed its median position.

LUNG: Electron microscope studies leave no doubt as to the existence of an alveolar epithelium in the adult and Low and Sampaio (1957), Anat. Rec., 127 show the continuity of the cubical and alveolar types. Boyden and Tompsett (1965) Acta anat. give beautiful models and injection preparations from premature and young infants.

VERTEBRAE: The account of the vertebrae, largely from Sensenig (1949) Carnegie Contributions to Embryology 23, replaces earlier versions which emphasized the fates of the parts of the scherotomes on either side of the sclerotomic fissure, the lighter cranial and the darker caudal sheets shown in our fig. The two sheets were said to move apart and the cranial sheet of one sclerotome to join the caudal sheet of the preceding sclerotome to build a vertebra, the so-called 'resegmentation' process. Sensenig and others have shown that there is no such movement. The position of the intersegmental artery opposite the body of the vertebra so glibly explained by the theory can be explained otherwise, as here.

UTERUS AND VAGINA: In this controversial field we follow Bulmer (1957), J. Anat., 91.

BRAIN: We have dropped the term 'diencephalon' and the terms related to it as they are used differently by different authors. Our description follows Bartelmez and Dekaban (1962), Contrib. Embryol., 37.

ROTATION OF GUT: We follow Lauge-Hansen (1960), The development and the embryonic anatomy of the human gastro-intestinal tract, Centrex, Eindhoven. A paper by a radiologist based on dissection and not, as is more usual, by an embryologist based on reconstruction models.

ANAL CANAL: A region technically difficult to study discussed by Tench, Amer. J.Anat., **59**, 333 (1936) and Moore and Lawrence, Surgery **32**, 352 (1952). The embryonic anal canal is often described as a 'proctodeum', suggesting the existence in the early embryo of a surface depression in relation to the end of the hindgut similar to the stomodeum which does develop in relation to the end of the foregut. There is such a proctodeum in echinoderms but not in man and we have avoided the term. Similarly the term 'anal folds' suggests something similar to the urethral folds, not the low ring-like elevation over the developing anal sphincter.

TERATOGENS: The effects of thalidomide and other teratogens are fully discussed in Goldstein, Aronow and Kalman, Principles of Drug Action, New York: Hoeber.

POSTMATURITY: Zhemkova and Topchieva (1964), Nature, **204**, 703. References to milk ridges, dermatomes, the ventral mesocardium and ventral mesogastrium are omitted, for lack of evidence of their existence in human development.

We have attempted to tell a straightforward story based on the evidence, rather than restate popular views even if we do not subscribe to them. But knowing so little and living so far away from the great libraries we have doubtless made many errors, and we would be grateful to anyone who could, in case there is another edition sometime, put us right.

INDEX

Abdomen, 159
Abdominal pregnancy, 153
Abdominal wall, ventral, 178, 186, 205
Abnormalities. *See* respective organs
Accessory air sinuses, 202, 234
Achondroplasia, 203
Acini, of glands, 117
Acoustico-vestibular ganglion, 59
Alar lamina, 57
Alar process, 105
Albinos, 217
Allantois, 11, 30, 188
Alveoli, pulmonary, 132
Amelia, 199
Ameloblasts, 114
Amnion, 10-12, 94, 148, 149, 154, 186
Amniotic ectoderm, 10
Anal canal, 181
Anencephaly, 232
Angioblasts, 33
Anococcygeal body, 30, 179
Anomalies, 240
Anomalies, of allantois, 188
  of anus, 182
  of aorta, 121, 144
  of arteries, 121, 144
  of breast, 196
  of cloaca, 183, 185
  of cranium, 203, 232
  of diaphragm, 166
  of genitalia, 185, 188, 193, 194, 196
  of heart, 144
  of hymen, 185
  of implantation, 152, 153
  of intestine, 168, 169, 171, 172, 215
  of kidney, 177, 180
  of limbs, 199, 200, 213
  of lip, 106
  of meninges, 210
  of nose, 106
  of neck, 126, 172
  of palate, 111
  of penis, 187, 193, 196
  of placenta, 143, 145, 146
  of spinal column, 172, 210
  of spinal cord, 171, 210, 217
  of umbilicus, 177, 188
  of urethra, 187, 193
  of urinary bladder, 187, 188
  of uterus, 187
Antrum, mastoid, 234
Anus, 181-183
Aorta, 33, 36, 118-121, 137, 143, 144
Aortic arches, 33, 53, 118-121
Aortic bodies, 174
Aortic sac, 53, 118, 120
Appendix vermiformis, 168
Aqueduct, cerebral, 55
Arachnoid mater, 228

Arnold-Chiari Malformation, 217
Arrector pili, 235
Arteries, axial, 77, 118, 212, 213
  carotid, 35, 118, 119
  coeliac, 65, 166
  coronary, 145
  hyaloid, 230
  hepatic, 166
  intersegmental, 35, 118
  intestinal, 35, 65, 166
  mesenteric, inferior, 65
  mesenteric, superior, 65
  middle sacral, 35
  of limbs, 212, 213
  pulmonary, 118
  renal, 176, 177
  subclavian, 118, 119
  umbilical, 34, 75
  vitelline, 35, 65
Atresia, of bile duct, 170
  of intestine, 169
  of oesophagus, 131
Atrial absorptions, 138, 139
Atrial septa, 45, 136
Atrioventricular bundle (of His), 140
Atrioventricular valves, 138
Atrium, left, 38, 41, 45, 136, 139
  right, 38, 41, 45, 136, 138
Auditory ossicles, 121, 233
Auditory placode. *See* Otic placode
Auditory tube, 126, 233
Auricle, of ear, 125
  of heart. *See* Atrium
Autonomic nerves, 214
Axial skeleton, 21, 207-211

Basal lamina, 57
Bile duct, 64, 170
Blastocyst, 7
Blood islands, 30
Bones, development of, 197-210
  of face, 201
  of limbs, 197-200
  of middle ear, 121, 122
  of skull, 200-203
  of trunk, 204-210
Brain, 23, 54, 217-226
Branchial arches, 51, 103, 118-124
Branchial cysts, 126
Branchial grooves, 50, 125
Branchial muscles, 122-124
Branchial nerves, 56, 59, 122, 224
Branchial pouches, 50, 126
Branchial skeleton, 121, 122
Breast, 195
Broad ligament, 195
Bronchus, 51, 132
Bulbar ridges, 137
Bulbus cordis, 137
Bursae, 212

Caecum, 65
Calyces, 175
Cardiac jelly, 39, 44
Cardiogenic plate, 29
Cartilages, laryngeal, 131
Cauda equina, 217
Cephalic flexure, 54
Cerebellum, 225
Cerebral aqueduct, 55, 225
Cerebral commissures, 222
Cerebral cortex, 220
Cerebral hemispheres, 54, 218
Cerebral ventricles, 54, 55, 218, 220, 223, 225
Cerebrospinal fluid, 226, 228
Cervical flexure, 225
Cervical sinus, 126
Choanae, 107
Chondrocranium, 200
Choriocarcinoma, 9
Chorion, 12
Chorionic villi, 13, 149-155
Choroid plexus, 220, 223, 225
Choroidal fissure, 48, 54, 230
Chromaffin bodies, 174
Ciliary body, 229
Ciliary ganglion, 214
Circulation, changes at birth, 142
  embryonic, 35
  foetal, 141
  intervillous, 13, 155
  umbilical, 34
  vitelline, 35
Circumnasal rim, 105
Cleft palate, 111
Cleidocranial dysostosis, 203
Clitoris, 181
Cloaca, 30, 76, 177
Cloacal membrane, 16, 30, 76, 181
Club foot, 213
Cloacal septum, 178
Coarctation of aorta, 143
Cochlea, 233
Coelom, communications of, 21, 94
  extraembryonic, 11, 149, 188
  intraembryonic, 21, 36, 194
Collecting tubules, 175
Coloboma, 230
Colon, 65, 170
Columella, 105
Commissures of brain, 222
Conjunctiva, 231
Connecting stalk, 11, 29
Cornea, 229
Coronary groove, 44
Coronary sinus, 139
Corpus callosum, 222
Corpus striatum, 218
Cotyledons, 153
Cranial nerves, 56, 59, 107, 115, 122-

124, 224, 231
Crista terminalis, 140
Cyclopia, 232
Cytotrophoblast, 9

Decidua, 9, 147
Dental lamina, 112
Dental papilla, 113, 114
Descent of ovary, 195
Descent of testis, 194
Determination, 77
Diaphragm, 164-166
Diaphragmatic hernia, 166
Diverticulum, hepatic, 62
  mammillary, 218
  Meckel's, 171
  olfactory, 218
  optic, 47
  pineal, 218
  respiratory, 51, 89, 132
  thyroid, 51, 128, 129
Duct, bile, 64, 170
  mesonephric/Wolffian, 71, 190, 191
  paramesonephric / Müllerian, 183-185, 188
  pronephric, 71
  salivary, 116
  thryoglossal, 128, 129
Ductus arteriosus, 119, 142, 144
Ductus deferens, 179
Ductus endolymphaticus, 232
Ductus venosus, 69
Duodenum, 65, 68, 160
Dura mater, 227, 228

Ear, drum, 234
  internal, 232
  middle, 233
  pinna, 114, 125
Ectoderm, 10
Ectopia vesicae, 187
Ectopic pregnancy, 152, 153
Efferent ductules, 190
Embedding, 9
Embryo, rotation of, 147
Embryoma, 158
Embryonic disc, 11, 15
Embryonic hormones, 48, 238
Embryonic induction, 236
Embryonic pole, 8
Embryotrophe, 9, 13
Enamel organ, 112, 114
Endocardial cushions, 45
Endoderm, 10
Endometrium, 7, 147
Endothelium, 30
Ependyma, 57
Epicardium, 39

Epididymis, 190
Epipexicardial ridge, 125
Epispadias, 187
Exocoelic membrane, 10
External acoustic meatus, 125
Eye, 48, 54, 84, 228-232
Eyelids, 231

Face, 103, 122-124
Faecal fistula, 171
Fallot's tetralogy, 144
Fertilization, 4
Floor plate, 57
Foetal circulation, 141-143
Foetal membranes, 154
Foetal movements, 211
Fontanelles, 202
Foramen, epiploic, 66
  interventricular, of brain, 55
  of heart, 43
  of Luschka, 226
  of Magendie, 226
  Foramen caecum, 115
  Foramen ovale, 136, 141, 144
Forebrain, 25, 54
Foregut, 27, 64
Fornix, 222
Frontonasal process, 104
Fourth ventricle, 52, 211
Froriep's ganglion, 60
Function, onset of, 240

Gall bladder, 62
Ganglia, autonomic, 214
  cerebrospinal, 56, 60
Genetic defects, 6
Genital swellings, 191
Genital tubercle, 178
Germ cells, primordial, 189
Germ layers, 9
German measles, 152, 228, 233
Globular process, 105
Gonads, 189
Greater omentum, 159
Gubernaculum, 194, 195
Gums, 112

Hair, 235
Hamartoma, 239
Hare lip, 106
Heart, 33-46, 88, 89, 134-146
Hepatic diverticulum, 62
Hermaphroditism, 196
Hernia, congenital inguinal, 194
  diaphragmatic, 166
  umbilical, 188
Hindbrain, 55
Hindgut, 27
Hydramnios, 132, 232
Hirschsprung's disease, 215

Hydrocele of cord, 180
Hydrocephaly, 232
Hymen, 184
Hyoid bone, 122
Hyoid operculum, 125
Hypophysis, 222
Hypospadias, 193
Hypothalamus, 219

Imperforate anus, 182
Imperforate hymen, 185
Implantation, 7
Incus, 121
Iniencephaly, 172
Inner cell mass, 8
Insula, 220
Intermediate cell mass, 32, 71
Internal capsule, 220
Intervenous tubercle, 139
Intervertrebral disc, 208
Intervillous spaces, 13
Interzone, of joint, 197, 198
Intestinal atresia, 169
Intestinal loop, 65, 166
Iris, 229
Isthmus of brain, 25, 55, 81

Jelly space, 31, 39
Joint, synovial, 211

Karyotypes, 5, 6
Kidney, 76, 175-177, 180
Klinefeter's syndrome, 6
Klippel-Feil syndrome, 172

Labia majora, 191, 195
Labia minora, 191
Labyrinth, 233
Lacrimal gland, 231
Lacrimal sac, 231
Lacunae, trophoblastic, 9
Lamina, alar, 57
  basal, 57
  dental, 112
  labio-gingival, 112
  terminalis, 24, 54, 218
Larynx, 52, 131
Lens, 48, 228
Lesser omentum, 163
Lesser sac, 66
Ligament, broad, 195
  coronary, 163
  falciform, 187
  of ovary, 180, 195
  of uterus, round, 195
Ligamentum arteriosum, 142
Limb bud, 76
Lingual swellings, 115
Lip, 105
Liver, 62, 90, 92, 161, 186

L

Lung, 51, 89, 132
Lymph, nodes, 146
  vessels, 146

Malleus, 121
Mammary gland, 195
Mammillary bodies, 218
Mandible, 122
Mandibular arch, 51, 115, 122
Mandibular process, 103
Mantle layer, 57
Marginal layer, 58
Mastoid antrum, 234
Maxillary process, 103
Meckel's cartilage, 122
Meckel's diverticulum, 171
Meconium, 181
Medulla oblongata, 225
Melanoblasts/Melanocytes, 217
Membrane, bucconasal, 107
  buccopharyngeal, 26, 28
  cloacal, 16, 30, 76, 178, 181
  pleuroperitoneal, 165
  stomodeal, 28
  tympanic, 234
Membranes, foetal, 154
Meninges, 228
Mesenchyme, 32
Mesentery, 65, 169
Mesocardium, 39
Mesocolon, 170
Mesoderm, primary, 10
  secondary, 17-19, 32
Mesogastrium, 66-68, 159
Mesonephric duct, 71, 190, 191
Mesonephric tubules, 72, 190, 191
Mesonephros, 73
Metanephric bud, 76
Metanephrogenic cap, 76
Microcephaly, 232
Microstoma, 114
Midbrain, 25, 55, 225
Middle ear, 233
Midgut, 27, 64
Migration, of kidney, 176
Milk line, 196
Mongolism, 6
Morula, 7
Muscles, arrector pili, 235
  branchial, 122, 123
  of eye, 206
  of head, 206
  of limb, 206
  of trunk, 204
Myelocoele, 210
Myocardium, 39, 43, 140
Myoepicardial mantle, 39
Myotomes, 204-206
Myocele, 204

Nail, 234
Naris, anterior, 104
  posterior, 107
Nasal capsule, 200
Nasal cavity, 103-109, 112
Nasal conchae, 112
Nasal field, 105
Nasal fin, 104, 105
Nasal pit, 103
Nasal placode, 47, 85
Nasal septum, 108, 112
Nasolacrimal duct, 231
Nasopharynx, 112
Neck, 26
Nephrogenic cord, 71, 76
Nerves, autonomic, 214
  cranial, 56, 59, 107, 115, 122-124, 224, 231
  spinal, 58
Neural arch, 197, 198
Neural crest, 55, 216
Neural folds, 23
Neural groove, 23
Neural plate, 22
Neural tube, 24
Neurenteric canal, 19, 171, 172
Neurilemma, 216
Neuropores, 24
Nose, 103-109
Notochord, 21
Notochordal canal, 17
Notochordal plate, 19
Notochordal process, 16
Nuclear columns, 224
Nucleus pulposus, 209

Oblique facial cleft, 231
Odontoblasts, 114
Oesophagus, 52, 130
Olfactory diverticulum, 218
Omentum, greater, 159
  lesser, 163
Optic chiasma, 231
Optic cup, 48, 54, 84
Optic diverticulum, 47
Optic groove, 25
Optic nerve, 231
Optic stalk, 48
Optic vesicle, 48
Oral membrane, 26, 28
Ossification, 198
Ostium primum, 46
Ostium secundum, 46,
Otic placode, 47
Otocyst, 49
Ovary, 190
  ligaments of, 195
Ovum, 4

Palate, 108, 110-112
Pancreas, 68, 92, 160
Paranasal sinuses, 202
Parathyroid glands, 128
Paraurethral ducts (of Skene), 193
Pelvic viscera of newborn, 185
Penis, 181, 187, 191-193
Pericardio-peritoneal canals, 36, 52
Pericardio-pleural fringe, 53
Pericardium, 23, 36, 164
Pharyngeal floor, 51, 129-132
Pharyngeal pouches, 50, 53, 125-130
Pharyngotympanic tube, 126, 233
Pharynx, 28, 50, 82
Philtrum, 105
Phocomelia, 199
Pia mater, 227
Pineal gland, 218
Pinna, 114, 125
Pituitary gland, 28, 222
Placenta, 149-155
Placenta praevia, 153
Placental barrier, 151
Pleura, 53
Pleuroperitoneal membrane, 165
Polycystic kidney, 176
Pons, 225
Pontine flexure, 225
Postmaturity, 240
Porencephaly, 221
Prepuce, 193
Primary loop of gut, 65
Primitive node (knot), 16
Primitive streak, 16
Primordial germ cells, 189
Processus vaginalis, 194
Prochordal mesenchyme, 28
Prochordal plate, 11
Pronephric duct, 71
Prostate, 193
Pseudohermaphroditism, 196
Pupil, 229
Pupillary membrane, 230
Purkinje fibres, 140

Rathke's pouch, 28
Rectum, 178
Recurrent laryngeal nerves, 124
Respiratory diverticulum, 51, 89, 132
Retina, 229
Roof plate, 57
Rotation of duodenum, 67
Rotation of gut, 168
Rotation of stomach, 67

Salivary gland, 116
Sclera, 229
Sclerotome, 207
Scrotum, 194
Semicircular canals, 232

Seminal vesicle, 193
Septum, cloacal (uro-rectal), 178
    intermedium, 45
    interventricular, 44, 137
    lucidum, 223
    nasal, 108, 112
    primum, 45
    secundum, 136
    transversum, 28, 29, 36, 62
Sinus venosus, 38-46, 69, 135, 138
Skull, 200-203
Somite, 25, 204
Spermatozoon, 3
Spina bifida, 210
Spinal cord, 23, 25, 57, 215, 217
Spinal ganglia, 56
Spleen, 159
Stapes, 122
Sternum, 205
Stomach, 65, 160
Subarachnoid space, 228
Subdural space, 228
Stomodeal membrane, 26, 28
Stomodeum, 26, 28, 104
Sulcus limitans, 57
Surfactant, 133
Suprarenal gland, 174
Sympodia, 200
Syncytiotrophoblast, 9
Synovial joint, 211

Tail bud, 29, 76
Tarsal glands, 231
Tectum, 25, 55
Teeth, 112
Tela choroidea, 223
Teratoma, 239
Testis, 190, 194
Tetralogy of Fallot, 144
Thalamus, 55
Thalidomide, 200
Thymus, 129
Thyroid gland, 51, 128, 129
Tongue, 115
Tongue muscles, 206
Tonsil, 127
Trachea, 51, 130
Tracheo-oesophageal communications, 131
Transverse sinus of pericardium, 39
Trigone of bladder, 179
Trophoblast, 8, 12
Trophoblastic lacunae, 13
Truncus arteriosus, 42, 119, 137
Tuber cinereum, 54
Tubercle, genital, 178
Tuberculum impar, 115
Tubules, collecting, 175
    renal, 175
Turner's syndrome, 6

Twins, 155-158

Ultimobranchial body, 128
Umbilical cord, 148, 149
Umbilical fistula, 171, 188
Umbilical hernia, 188
Umbilicus, 188
Urachus, 188
Ureter, 76, 179
Ureteric bud, 76
Urethra, female, 179
   male, 179, 191-193
Urethral folds, 192
Urinary bladder, 179
Urinary fistula, 188
Urogenital sinus, 178, 181, 185, 191
Uro-rectal septum, 178
Uterine tube, 183, 195
Uterus, 184, 187
Uvula, 110

Vagina, 184
Vaginal process, 194
Vas deferens, 71, 190
Valves, cardiac, 46, 137, 138
Vein, azygos, 74, 135
   cardinal, common, 35, 134
   cardinal, post- 35, 73, 134, 135
   cardinal, pre- 35, 134
   cardinal, sub- 74
   coronary (sinus), 139

hepatic, 162
innominate, 134
internal jugular, 134
intersegmental, 35
oblique, of left atrium, 139
of limbs, 213
portal, 68, 160
pulmonary, 45, 139
subclavian, 134
umbilical, 33, 35, 69, 141, 149
vitelline, 33, 35, 68, 74, 135
Vena cava, inferior, 74, 135
   superior, 134
Ventricular loop, 40
Vermiform appendix, 168
Vertebrae, 207-210
Vesicle, ectodermal, 11
   endodermal, 11
   trophoblastic, 7
Villi, chorionic, 13, 149-155
Vitelline artery, 35
Vitreous body, 229
Vulva, 185

Wharton's jelly, 63

Yolk sac, 12, 27, 64
Yolk stalk, 27, 63

Zona pellucida, 4
Zygote, 5

Holmes McDougall Ltd., Perth.